An Observer of Observatories

The Journal of Thomas Bugge's
Tour of Germany, Holland and England in 1777

Edited by Kurt Møller Pedersen and Peter de Clercq

Aarhus Universitetsforlag

An Observer of Observatories
The Journal of Thomas Bugge's Tour of Germany, Holland
and England in 1777

© The authors and Aarhus University Press, 2010

Graphic production: P.J. Schmidt A/S
Cover: Jørgen Sparre
Illustration: "View of the Royal Observatory,
Greenwich", water colour, c. 1770.
© National Maritime Museum, Greenwich, London.
Printed in Denmark 2010

ISBN 978 87 7934 311 5

Aarhus University Press
Langelandsgade 177
DK-8200 Aarhus N
Fax: +45 89 42 53 80
www.unipress.dk

White Cross Mills
Hightown, Lancaster, LA1 4XS
United Kingdom
www.gazellebookservices.co.uk

PO Box 511
Oakville, CT 06779
www.oxbowbook.com

This book is published with the financial support of
The Aarhus University Research Foundation
The Carlsberg Foundation

Foliation in the journal manuscript is indicated as [3 recto].
A scan of the original manuscript is accessible at
www.kb.dk/permalink/2006/manus/659/dan.

Bugge often added words in the margin. These have been
inserted in the text between * *.

Contents

Preface

When Thomas Bugge (1740-1815) was appointed professor of mathematics and astronomy at the University of Copenhagen in January 1777, he became responsible for the observatory, which in the seventeenth century had been built on top of the Round Tower. Later that year he travelled to Holland and England to acquaint himself with the state of astronomy and instrument-making in these countries. After his return four months later, he began to renovate the observatory.

During his tour he kept a journal in which he noted what he saw, whom he met and which books and instruments he bought. It comprises five quarto notebooks with a total of 94 folios, filled on both sides with text and drawings. In 1969, this journal was discovered in the Royal Library in Copenhagen by the first editor of the present volume, associate professor Kurt Møller Pedersen. He brought it to the notice of Cdr Derek Howse (1919–1998), then head of the department of Navigation and Astronomy at the National Maritime Museum, Greenwich. Howse flew to Denmark especially to study the manuscript and judged it of such historical interest that he advised publication. In 1975, on the 300th anniversary of the foundation of the Royal Observatory, Greenwich, Bugge's journal was exhibited there for several weeks and received much attention from the hundreds of historians of science gathered for this important event.

This positive response was encouraging for Pedersen. Together with his secretary, Mette Dybdahl, he made a transcription of the Danish manuscript and an English translation. Soon copies began to circulate among scholars, who quoted and used it in their publications. In 1997, a modestly produced edition was issued by the History of Science Department of the University of Aarhus, entitled: *Thomas Bugge, Journal of a Voyage through Holland and England, 1777*. This was emphatically no more than a preliminary edition, and in his preface Pedersen expressed the hope that *'someone more qualified than I will undertake to improve the English translation and provide annotations of the instruments described and drawn by Bugge, so that a proper scholarly edition of this valuable source for 18th century astronomy may eventually be available.'*

In 2001 the second editor, Dr Peter de Clercq, came forward and offered his assistance. He began to make textual improvements and annotation, a prolonged process that at times came to a standstill as there was no clear prospect of publication. This editorial work was greatly facilitated when, in 2006, the Royal Library in Copenhagen made a scan of the manuscript available on its website: www.kb.dk/permalink/2006/manus/659/dan.

In the spring of 2008, Aarhus University Press agreed to publish the book, on condition that financial support would be found. Proposals were sent to the Aarhus University Foundation and the Carlsberg Foundation. It was a great moment for the editors when, in the autumn of 2008, both funds agreed to supply the necessary funding, and they deserve gratitude for this contribution to a better understanding of the history of science in Europe.

The Carlsberg Foundation supplied additional funding to allow the book to appear in two separate, but essentially identical editions: one in Danish, the other in English. This meant significantly more work for the editors, but it was undertaken with pleasure as, forty years after its discovery, Bugge's journal will now be available both in transcription and in an English translation, each fully annotated.

Many friends and colleagues, librarians, archivists, museum curators and others have contributed in a variety of ways. For supplying information and photographs, for commenting on parts of the edition, and for supporting the application for the funding of this publication, we wish to thank Jørgen From Andersen, Martin Beech, Jim Bennett, Jonathan Betts, Dan Charly Christensen, Gloria Clifton, Tiemen Cocquyt, Diana Crawforth-Hitchins, J.Th. van Doesburg, Rob van Gent, Willem Hackmann, A.J.E. Harmsen, Helge Kragh, Leif Kahl Kristensen, Peter Louwman, Anita McConnell, Alison Morrison-Low, Joshua Nall, Keld Nielsen, Erling Poulsen, David Riches, Sara Schechner, Bruno Svindborg, David Thompson, Anthony Turner, Jan van Wandelen, Jane Wess, Diederick Wildeman and Huib Zuidervaart, as well as those we may have failed to mention.

Finally, we thank the librarian at the Department of Science Studies, Susanne Nørskov and Aarhus University Press, and in particular Claes Hvidbak and Sanne Lind Hansen, for their expert handling of this project.

Aarhus and London, Autumn 2009
Kurt Møller Pedersen and Peter de Clercq

Thomas Bugge 1740–1815

Some forty years ago, a biography of Bugge was published, written by the Director of the Geodetic Institute to commemorate the 150th anniversary of his death.[1] More recently, a large amount of information on Bugge, especially on his astronomical activities, became available in a three-volume overview of four centuries of astronomy in Denmark.[2] As both publications are in Danish, those unable to read that language must look elsewhere for information on Bugge. They may turn to a commemorative volume on the Royal Danish Academy of Sciences and Letters, which he had served as its secretary for many years.[3] There is the introduction to a modern edition of his journal on the Parisian scientific scene of 1798.[4] And there are a few pages on Bugge in a recent book on the history of science in Denmark, which characterizes him as 'one of the most famous scientists of his time', and states that 'Thomas Bugge did not produce any scientific work of similar originality, but his overall contribution to the natural sciences in Denmark was far greater than Wessel's was'.[5] It makes one wonder why he was not included in the authoritative, multi-volume *Dictionary of Scientific Biography*.[6]

Thomas Bugge was born in Copenhagen on 12 October 1740 as the son of Peder Bugge and his wife Olive, born Saur. His father was a scribe taking care of the financial accounts of the royal wine cellar, and was later appointed chamberlain and superintendent of the royal household. They belonged to a line of a noble family but were unable to live as such because of insufficient means. At the age of sixteen, Thomas enrolled at the University of Copenhagen to study theology, graduating in 1759. During these years, he also studied pure and applied mathematics, supervised by Professor Christen Hee (1728–1782). He worked as an assistant to Christian Horrebow (1718–1776), the director of the observatory whom he was later to succeed. In 1761, he was sent to Trondheim in Norway to observe the Venus transit.

Bugge is primarily known for his work as director of the survey of Denmark, a project under the supervision of a special surveying commission of the Royal Academy of Sciences. At the age of twenty-two, Bugge became one of its first geographic surveyors. He immediately began surveying and lecturing on the subject and quickly became its driving force.[7] He was inspired by the

1   Andersen 1968

2   Thykier 1990. The sections most relevant to Bugge are vol. 1, pp. 94–104 (on his directorate of the observatory), vol. 2, pp. 184–187 (instrumentation of the observatory during his directorate) and 214–215 (line-drawings of the observatory and the main instruments in Bugge's time, based on Bugge 1784) and vol. 3, pp. 445–458 (including a list of Bugge's publications, which incidentally is not complete). There is an English summary on pp. 583–589.

3   Pedersen 1992

4   Crosland 1969

5   Kragh and others, pp. 151–154. The comparison is to the surveyor and mathematician Caspar Wessel (1745–1818).

6   Published in the 1970s under the auspices of the American Council of Learned Societies in fourteen volumes, with later supplements. It *does* have an entry on Wessel (vol. XIV, pp. 279-281).

7   He published a book on the theory and methodology behind the project, whose title translates as 'A Description of the Surveying Method used for the Danish Geographical Maps'; it was later also translated into German. Bugge 1779a.

Portrait at the observatory in Copenhagen of Thomas Bugge. Artist unknown.

Title page of Thomas Bugge's description of the Danish surveying project, here from the German edition 1787.

French surveying activities directed by César-François Cassini de Thury (1714–1784) who produced 182 maps of France on a scale of 1:86,400. Bugge used 1 metric inch on his plane table = 2000 feet in the field or 1:20,000. The topographical maps drawn on the plane table were then put together, engraved and published as 1:80,000, 1:120,000 and even 1:320,000. Bugge avoided Cassini's skew numbers like 1:86.400 by using a more practical decimal system that made calculations easier.[8] This work involved advanced astronomical observations and many trigonometrical calculations. In 1780 Bugge became de facto leader of the geographical, trigonometric and economic survey of Denmark. It was important to provide the military with reliable maps. It was equally important for a new land register, as the open fields, in which the land was communally owned by all farmers in a village, were being divided into individual fields. Bugge was also engaged in setting up methods for evaluating the site quality of each single field. The organization, teaching of surveyors, map production and funding must have consumed much of Bugge's time and energy. The project was much criticized after his death, but he was nevertheless the man who accomplished the country's first modern survey. It resulted in an immensely improved topographical knowledge of Denmark and generated an expertise in the field of surveying that had not previously existed.

In 1773, Bugge also became president of the Royal Agricultural Society (Det Kongelige Landhusholdningsselskab), which he was to remain for ten years. The second half of the eighteenth century saw in many European countries a growing interest in stimulating farming productivity, and the Royal Agricultural Society played an important role in proposing new strategies. His continuous interest in agricultural matters is evident from his journal. When he passed through Holstein on his way to Holland and England, he carefully noted how farms were constructed and how the mediaeval open fields were divided among the farmers, leading to a revolution in the farming system.

8   More about the decimal system used during the
     surveying of Denmark is found in Kristensen 2001.

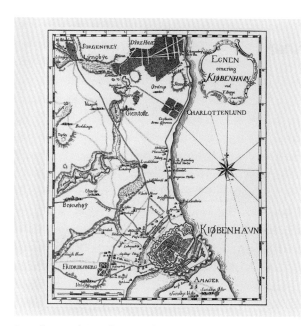

Bugge's map of Copenhagen and its environs 1766.

The original copperplate, 20.5 × 29 cm, is now with hundreds other plates at the Steno Museum in Aarhus.

On 30 January 1777, Bugge was appointed professor of mathematics and astronomy at the university and director of the observatory, situated on top of the Round Tower. This striking building, 36 meters high, with a spiral stairway wide enough for a horse-drawn carriage, had been erected as a prestige project at the behest of King Christian IV and had been completed in 1642. The very next year, the tower had been fitted with a telescope, which makes it one of the very first university observatories in the world, predating the observatories in Paris and Greenwich. However, its location in the heart of Copenhagen was not ideal, a far cry from the quiet of the nearby island of Hven, where Tycho Brahe (1546-1601) had built his Uraniborg. Before Bugge, a succession of Danish astronomers had used the Round Tower observatory, including Longomontanus (1562-1647), Ole Rømer (1644-1710), who also had a private observatory outside the city), Peder Nielsen Horrebow (1679-1764) and finally his son and successor, Christian Horrebow (1718-1776).

The King donated 7,000 rigsdaler to a renovation of the observatory and acquisition of new and better instruments.[9] This was not a small sum; the surveyors working in the field for the Danish mapping project were each paid 300 rdl. per year. But it was not a luxury budget either. Building and equipping the new Radcliffe observatory in Oxford, which Bugge visited, cost more than twenty times that amount.[10] Admittedly, the two projects are not entirely comparable. In Oxford, a large new building was erected, including a private home attached to it for the astronomy professor. Bugge only had to erect a shed on top of the Round Tower. Perhaps more to point is a comparison with what had been spent on the five main instruments, that the eminent London instrument-maker John Bird (1709-1776) had supplied to the Radcliffe observatory. His two mural quadrants, transit instrument,

9   Bugge 1784, p. XXVII, § 14.

10   Guest 1991 gives two different totals for the expenses on building and equipping of the Radcliffe observatory in the period 1773 to 1799 (when it was finally completed): £ 31,661 (p. 246) and £ 35,750 (p. 508, note 1). Either way, this is incomparably more than the 7,000 rigsdaler, which equal £ 1400. On the equation 1 pound = 5 rigsdaler, see below, page xxiii, A note about money.

Bugge's observatory on top of the Round Tower seen from east, c. 1780. The Royal Library, Copenhagen.

zenith sector and equatorial sector had cost almost exactly as much as the entire royal donation to modernize the Copenhagen observatory. [11]

When he published his observations six years later, Bugge preceded his tables with a detailed description of the renovated observatory and its instruments, illustrated with fine engravings.[12] An octagonal room of 25 feet diameter housed a 6-foot mural quadrant and a 12-foot sector, and was flanked by two smaller rectangular rooms, the western one housing a 6-foot transit instrument, the eastern one a 4-foot diameter portable astronomical circle. The last angular measurement instrument that Bugge mentioned is a portable 3-foot quadrant on a tripod. All five had been constructed by the instrument-maker Johannes Ahl (1729–1795), who had been an apprentice and later a partner of the eminent Swedish instrument-maker Daniel Ekström (1711–1755), but had then left Sweden and set up a workshop in Copenhagen. Ahl had already worked for the

mapping project, and in his journal Bugge refers to 'the Danish geographical instrument' made in 1762 that he had used in the survey. There were also time-keepers by Mudge and Dutton from London, Jahnson from Copenhagen and Le Paute from Paris, and telescopes by Dollond from London and other makers. In later years, Bugge would also acquire telescopes by among others William Herschel and Nairne and Blunt.[13]

But in spite of his efforts to reorganize and modernize the observatory, Bugge did not bring astronomy in Denmark to a very high international standard. He studied Algol and found that it rotates in 2 days, 21 hours and 7 minutes, and that Saturn rotates in 6 hours, 4 minutes (the modern value is approximately ten hours). His interest in geomagnetic theories also led him to make systematic measurements of magnetic intensities. He contributed four papers to the *Philosophical Transactions of the Royal Society*, but many of his findings never found their way into international journals. However, Bugge communicated many of his results in private letters to scientists all over Europe. This correspondence is now in the Royal Library in Copenhagen as a bequest by Bugge – which also included the manuscript of his travel journal – and deserves to be studied to find out more about his scientific work.[14]

Ardent and competent as he was, he also became correspondent for the Societas Meteorolog-

11   These five instruments had cost £1392, see Morrison-Low 2007, p. 138.

12   Bugge 1784. In the British Library are two presentation copies (shelf-marks 50i3 and 434g1) with Bugge's handwritten dedications. One is to King George III, whose private observatory at Richmond he had visited; the other to Sir Joseph Banks, the President of the Royal Society.

13   Of these, only the portable quadrant (Round Tower Museum), and the 1762 geographical instrument (Kroppedal Museum) survive. Thykier 1990, pp. 186–187.

14   There are 84 letters from and to Bugge dated between 4 February 1779 and 6 April 1814 in Royal Library NKS 287, 1304 and 2749. The correspondents are Aeneae, Arnold, Banks, Dryander, Herschel, Hoppe, Hornsby, Huygens, Jones, Mackenzie, Magellan, van Marum, Maskelyne, Nairne (& Blunt), Parker, Phillips, Ramsden, Royal Society, van Swinden, Taylor and Wolff & Dorville. Based on a list drawn up by Jørgen From Andersen, curator of Hauchs Physiske Cabinet, Sorø. For other surviving correspondence with Banks, Bidstrup and Herschel, see Morisson-Low 2007, pp. 154–158.

ica Palatina, an international network for weather observations, organized from Mannheim.[15] Bugge listed the instruments that had been supplied for the purpose, and published his weather observations.[16]

Bugge was rector of the university for three periods (1789–90, 1801–02 and 1810–11) as well as secretary of the Royal Academy of Sciences from 1801 until his death. He was a member of many academies throughout Europe. In 1788 he was elected a Fellow of the Royal Society, and among those who backed him were four men whom he had met during his visit to England a decade earlier: the astronomers Nevil Maskelyne, Thomas Hornsby and Anthony Shepherd, and the instrument-maker Jesse Ramsden.[17]

In 1798, Bugge went to Paris as Danish representative at the international conference for the introduction of the metric system of weights and measures. He was there for six months but, frustrated with the endless procrastination, left before the conference actually started.[18]

In September 1807 Bugge suffered a blow when the English navy bombarded Copenhagen in an attempt to stop Denmark from joining Napoleon's Continental System. Bugge's house was hit by 35 bombs and burnt down. He lost all his furniture and goods, his library of 7000 volumes, his collection af mathematical and physical instruments, his maps, in short: The fruits of a lifetimes work.

Bugge died on 15 January 1815 and was buried in the Assistens Cemetery in Copenhagen. When a member of the Royal Academy of Science died, it was customary that a member would assess his achievements in a public speech. This did not happen in Bugge's case, possibly because of his strained relationship with the man who succeeded him as secretary of the Academy. This was Hans Christian Ørsted (1777–1851), one of Denmark's most celebrated physicists. He was a generation younger, having been born on the day Bugge arrived in Bremen on his way to Holland and England. He graduated as a pharmacist in 1797, and received his doctor's degree in 1799 for a dissertation on Immanuel Kant's philosophy. He travelled in Germany in 1801–02, where he was much taken by the romantic philosophy, which was to some extent opposed to a Newtonian-Laplacian philosophy that was more to Bugge's liking.[19] When Ørsted applied for a position as professor of physics, he was not successful, and Bugge in particular seems to have had his reservations. The two men were never thereafter on good terms. Ørsted was appointed extraordinary professor, and became ordinary professor of physics only in 1817. Three years later he made his discovery of the magnetic effect produced by an electric current, which was to open up a new chapter in pure and applied physics. In a way it is regrettable that Bugge, who himself had contributed so much to the natural sciences in his country, could not witness this epochmaking moment.

Travel journals

Travel journals are a valuable source of information for historians, and this includes those with a special interest in science and scientific instruments, to whom Bugge's journal will probably appeal most. Through the traveller's eyes, we see

---

15   Cassidy 1985; for Bugge's participation from 1782 to 1788, see pp. 23–4.

16   Bugge 1784, § 48, lists a barometer, a thermometer, a hygrometer and a declinatorium, as well as two corresponding hygrometers made by Baron de Gedda, and an anemometer after the invention of Wilcke (which Bugge considered unreliable). His observations are § 79.

17   The original document, signed in April 1787, cat. no. EC/1787/21, can be seen at http://royalsociety.org, section Library and Archives. It was signed by nine men who proposed to accept his candidacy, the others were William Wales, George Shuckburgh, Charles Blagden, John Sinclair and William Watson.

18   Alder 2002, pp. 259–260

19   In his 1796 textbook *De Første Grunde til den Sfæriske og Theoretiske Astronomi* ('First principles of spherical and theoretical astronomy'), Bugge wrote in § 113 that 'Newton's ingenious System' had changed from being a hypothesis to 'being a mathematical certainty.' The Bugge–Ørsted relation is decribed in Christensen 2009.

universities, academies and societies in action, we step into observatories and laboratories, into the homes of scholars and collectors, and into the workshops of makers of instruments, clocks and watches. While undoubtedly many more such documents lie unpublished in private and public archives, a fair number have been made available to researchers. To place Bugge's manuscript in perspective, we shall discuss some of them.

The notes that the German book collector Zacharias Conrad von Uffenbach (1683–1734) made on his intellectual foray through northwestern Europe in 1709–11 were published twenty years after his death.[20] Among others, they contain uniquely detailed comments on his visit to the Leiden instrument-maker Jan van Musschenbroek.[21] Later English editions of the sections in which he discusses his visits to Oxford and London[22] are often quoted by instrument historians.

Six journals written by the Dutch scholar Martinus van Marum on his foreign travels undertaken between 1782 and 1802 have also been published. They shed light on his acquisition of fossils, geological specimens and instruments for the recently founded Teyler's Foundation in Haarlem. We single out the two dealing with Europe's metropoles. In 1785 he was in Paris where he met Antoine-Laurent Lavoisier (1743–1794) and Benjamin Franklin (1706–1790), and presented the Académie des Sciences with a copy of the printed description of his new electrical

machine. In 1790, he was in London, where he bought instruments from among others Edward Nairne and George Adams, both of whom Bugge had patronized thirteen years earlier.[23]

If Bugge's journal is exceptional in that it contains many sketches and drawings, we know one other published travel journal that is also profusely illustrated by its author. This is the diary that a Swedish iron master, Reinhold Rücker Angerstein (1718–1760), kept of an industrial espionage trip through Britain in the 1750s. He gathered information on mines, factories and foundries in the industral regions of England and Wales. Historians of science will be interested to find that in London, Angerstein attended a course of physics demonstrations that the lecturer on natural philosophy, Erasmus King, gave in his 'Experiment room'.[24] It involved among others that emblematic device of 18th-century physics, the air-pump. A quarter of a century later, Bugge would buy one for himself in London, with a complete set of accessories for experiments. This serves to underline that the second half of the 18th century was still a time before the great specialization. Bugge was appointed professor of mathematics and astronomy, but he also had a practical interest in experimental physics.

Having said that, the main object of this tour was of course to study observatories and their instruments. We therefore conclude with the travel accounts of three other astronomers who visited Holland and/or England in roughly the same period.

Bengt Ferrner (1724–1802), professor of astronomy at Uppsala, was employed as a compan-

20   Uffenbach 1753–54. This edition was prepared by the clergyman and librarian Johann Georg Schelhorn, to whom von Uffenbach had left his papers.

21   Uffenbach 1753–54, pp. 430–437; de Clercq 1997a, pp. 53-54, 109, 118, 153 and 220-222.

22   Quarrell and Quarrell 1928; Quarrell and Mare 1934.

23   The journals were published in the second volume of Forbes e.a. (1969–1976). For a discussion of his trip to London, see Levere 1973, pp. 54–64. His machine, the largest plate electrostatic generator ever produced, and the instruments bought in London, survive in Teyler's Museum in Haarlem and are described and illustrated in Turner 1973.

24   Angerstein 2001; his (brief) reports on the lectures are on pp. 21-25. On Erasmus King and other self-employed lecturers in mid-18th century London, see Morton and Wess 1993, pp. 67–87.

◀ *The British bombardment of Copenhagen 1807. C.W.Eckerberg's drawing shows the burning of the church of Our Lady behind the Round Tower. The Royal Library, Copenhagen.*

ion-tutor to the son of a wealthy merchant. In the years 1758 to 1762, he took his pupil on an educational tour through Europe which, like Angerstein's trip, had an element of industrial espionage. Ferrner kept a detailed journal, which survives in the Royal Library in Stockholm. It was published in 1956 in a fine edition, but in Swedish and therefore inaccessible to most researchers.[25] Only the section dealing with his visits in western England have appeared in English.[26] Dutch historians are fortunate to have the section dealing with Ferrner's stay in Holland available in a Dutch edition.[27] Like Bugge twenty years later, Ferrner visited the observatory at Leiden University and Jacobus van de Wall's private observatory in Amsterdam.[28] In England, he spent several months in London and made brief visits to Cambridge and Oxford. He records his contacts with researchers, instrument-makers and clock-makers, including the specialist maker of reflecting telescopes, James Short (1710–1768) and the maker, among much else, of the mural quadrants for observatories, John Bird (1709–1776). He attended meetings at the Royal Society, and saw the observatory at Greenwich and that of the Earl of Macclesfield outside Oxford.

A more prominent astronomer, who kept journals of his foreign visits, is the Frenchman Joseph-Jérôme le Français de Lalande (1732–1807). We have no evidence that Bugge knew Lalande personally, but he certainly knew him as an author. In a bookshop in the Strand, he bought Lalande's Ephemerides (printed tables showing the daily positions of heavenly bodies), and, as one would expect, he had read his well-known text-book As-

tronomie.[29] The diary of Lalande's visit to England in 1763 was published in 1980, and more recently an English version was made available on-line.[30] Lalande's diary is written more in telegraphese than Bugge's, but is comparable in content. He reports on meetings with some of the same scholars that Bugge met, such as Maskelyne at Greenwich and Hornsby at Oxford. He also met London's most prominent instrument-makers, including the afore-mentioned Short and Bird, who would both be dead by the time Bugge visited the capital. He also showed sufficient interest in technical and industrial matters to take in some of the same sights that Bugge would visit, such as 'the fire-pump' at Chelsea, the gun-foundry at Woolwich and the collection of models and machines at the Society of Arts.

In 1774, Lalande travelled to Holland in connection with the launch of the Dutch edition of his astronomy text-book. His attempt to stimulate a greater role for astronomy in navigation met with such a lukewarm response, that he came away with a low esteem of the state of astronomy in the Low Countries. On this visit too, Lalande kept a travel diary, but this has not yet been published.[31]

A third astronomer whose travel accounts we have is the Swiss-born Jean Bernoulli (1744–1807). At the age of nineteen he had been appointed Astronomer Royal at Berlin. In his mid-twenties, he travelled widely and published his impressions, in the form of fourteen fictional letters, in a French book whose title translates as *Astronomical Letters. In which we give an idea of the*

25 Ferrner 1956.

26 Woolrich 1986.

27 Kernkamp 1910.

28 Zuidervaart 1999 discusses Ferrner's visits in Holland, which also included the observatory at Utrecht University (which he found 'pityful') and meetings with two astronomers, the competent Nicolaas Struyck (1686–1769) and the dabbler Pieter Gabry (1715–1770).

29 See Bugge's journal, appendix 2 and folio 70 verso.

30 The English version, Watkins 2002, contains far more extensive comments on Lalande and his diary than the French edition, Lalande 1980.

31 Zuidervaart 1999, p. 333, shows that in his publications Lalande gave an undeservedly negative image of the state of astronomy in Holland, which was to persist throughout the 19th century. Lalande's manuscript is in the Bibliothèque de l'Institut de France (Paris), Cote: MS 2195. Huib Zuidervaart intends to prepare an annotated edition in due course.

present state of practical astronomy in several European towns in Europe.[32] Bugge knew the book, and probably even had it with him on his tour.[33] Bernoulli gave detailed descriptions of the observatories and their main instruments that he visited in Kassel, Frankfurt, Paris, Strasbourg, Basle and – most relevant in our context – Greenwich, Oxford and Cambridge. Instrument historians often quote Bernoulli's enthusiastic report on the amazing riches of the London instrument shops. He attended the sale of the instruments left by the eminent telescope maker James Short, and for some noted the prices fetched and the names of the buyers. Among them was the merchant and amateur astronomer William Russell, whom Bugge would visit eight years later.[34]

All these accounts by 18th-century travelling scientists are of great interest. But it seems fair to say that none of them recorded their visits to the observatories or to some of the leading instrument-makers and clock-makers in quite the same detail as Bugge did. Perhaps he was keener than any of the others to learn as much as he could, hoping to apply it to the improvement of his own work-place.

Bugge's journal of 1777

In the forty years since its discovery, Bugge's journal has been used by researchers in and outside Denmark. In the 1970s, a provisional and incomplete translation became available informally to a few British researchers, and tantalizing references to the journal appeared in their publications. Examples are Turner's study on Martinus van Marum's instruments in Teyler's

Museum,[35] Howse's book on the buildings and the instrumentation of Greenwich Observatory,[36] and Millburn's study on the London instrument-maker Benjamin Martin.[37] The preliminary edition of 1997 brought the journal into wider use among historians of the scientific instrument trade. Examples are an unpublished conference paper,[38] and studies on the instrument-maker George Adams,[39] on instrument shops in London and Paris,[40] and on the English instrument-making industry in the Industrial Revolution.[41] The Dutch leg of the tour has received less attention, although one scholar has made good use of Bugge's notes on the Amsterdam amateur astronomer Jacobus van de Wall.[42] The editors of the present volume themselves have also written short articles drawing attention to the journal.[43]

The main ingredients in Bugge's journal, which define its value as a source of information for historians, are his contacts with the learned community in Holland and England and his visits to the work-places of science. He met university professors, amateurs astronomers and makers of instruments and precision time-keepers. He inspected astronomical observatories and university collections, saw the shops and workshops of mechanics, and witnessed physical experiments.

32 Bernoulli 1771. We are not aware of any English edition of this publication.

33 See his comment on a telescope in the observatory at Christ's College, Cambridge, folio 77 recto: "a telescope with the device for motus parallacticus which is described by Bernoulli in his Lettres Astronomiques, pag. 118–119".

34 See de Clercq 2007, pp. 30–31 and de Clercq 2009, pp. 27–28. A transcript of the auction sale catalogue is accessible at www.mhs.ox.ac.uk/library/ephemera

35 Turner 1973a, pp. 22 and 36.

36 Howse 1975, pp. 37, 114 and 148 and fig. 107.

37 Millburn 1976, pp. 164-165.

38 Gloria Clifton, 'Thomas Bugge in England: A Danish view of London Instrument Making in 1777', unpublished paper read at the 18th Scientific Instrument Symposium at Sorø Academy, July 1998.

39 Millburn 2000, p. 193.

40 Bennett 2002.

41 Morrison-Low 2007, uses Bugge's journal in a section named 'How Foreigners Saw the London Trade', pp. 150-162, which also discusses the career of Bugge's protégé, the instrument maker Jesper Bidstrup.

42 Zuidervaart 1999, pp. 312–313; Zuidervaart 2003; Zuidervaart 2004, pp. 426, 435–436 and 443, note 137.

43 Pedersen 1982, Pedersen 2001 and de Clercq 2005 (the latter with a focus on the weeks that Bugge spent in the Netherlands).

As noted, historians have already begun to use the journal in their researches, but there remain rich pickings for specialists. A case in point are Bugge's notes on horological matters, which shed new light on various issues. During his visit to the precision clockmaker John Arnold (1735/36–1799), he describes and draws the double T-balance, which shows that Arnold made this earlier than had previously been realised. His notes also confirm that it was Arnold who invented the five bar gridiron, using a zinc alloy, something that had been suspected but not before confirmed. Later, in Aubert's private observatory, Bugge describes a regulator by the London clockmaker John Shelton, and specifies – again with a drawing – that *"every 5 seconds have been marked with a long mark in order to avoid miscounts"*. This is the earliest known reference to the five-second marks and shows that they were originally on Shelton's dials, whereas some have suggested that they were all put on later.[44] Undoubtedly, other specialists will find similar material of interest to them.

Bugge visited eleven observatories, two in Holland and nine in England, and gives a wealth of information on his discussions with the astronomers he met there, and on the buildings and the instruments, with many detailed drawings. Of course they included the one at Leiden, the oldest university observatory in Europe, and the two great ones in England, the Royal Observatory at Greenwich and the new Radcliffe Observatory at Oxford. But he also gives detailed information on less well-documented, private observatories, such as that of the merchant Alexander Aubert at Deptford, just outside London, which he called *"the most complete in Europe for its size"*. In these observatories and in the workshops of the instrument-makers and clockmakers, Bugge learned much about instrumentation and about practical solutions. He brought these ideas back to Copenhagen, where he instructed Johannes Ahl to incorporate them in the instruments that he made for the Round Tower. A case in point is the clever lighting device on the transit instrument, which Bugge had seen in the Oxford observatory.[45]

One may wonder why he did not order them from London which, as he had seen with his own eyes, was the undisputed centre of precision instrument-making at the time. The eminent maker John Bird was no longer alive, but he could have turned to the new rising star for large observatory instruments, Jesse Ramsden, whom he had met. He probably realized that that would be too expensive; as we have seen, he had only a limited budget. But perhaps more importantly, Bugge hoped to stimulate the manufacture of instruments in his own country. Employing Johannes Ahl in the modernization of the observatory was therefore a logical decision.

But he did not return from London empty-handed. As documented in his journal, he bought some small mathematical and optical instruments, and a set of apparatus for physics experiments, including an airpump and an electrical machine, with all accessories for demonstrations, presumably to be used in his university teaching.

Technology transfer through industrial espionage is a recurrent theme in the history of industry and technology, and we know of Scandinavians who tried to get a 'look behind the scenes' of leading London instrument-making workshops. In 1787, Bugge himself was to send the aspiring Danish instrument-maker Jesper Bidstrup (1763–1802) to London to learn the trade, and this included some spying.[46] And two years later, the Swede Jøns Matthias Ljungberg (1748–1812) 'bribed' a man working for Jesse Ramsden to reveal details of the tube-drawing machine employed in the

44 These three examples of novelties found in Bugge's journal (45 recto, 49 recto and 85 recto) are chosen from detailed comments on the horological sections in the document, kindly supplied by Jonathan Betts, Senior Specialist Horology at the Royal Observatory, National Maritime Museum, Greenwich.

45 This and other examples are given in Pedersen 2001.

46 Christensen 1993, and Morisson-Low 2007, pp. 154–160.

workshop, complete with a detailed drawing.[47] But none of this applied to Bugge himself, who as an academic and a potential customer was received with great openness by the instrument-makers and clockmakers that he visited.

It would be wrong to think that Bugge had only eye for observatories and workshops, and that his journal is only of interest to historians of science, instruments and time-keepers. Bugge travelled with his eyes wide open and noted much else besides. Like any other tourist, he took in the sights, commenting on public buildings and popular entertainments. When he visits churches, he shows a dislike of gothic elements; not for him the renewed appreciation of medieval art and architecture that we now refer to as the Gothic revival. He could be prudish: the cafes in a popular area in Amsterdam *seemed more like a brothel than a decent and cheerful pleasure.* He had sufficient interest in the arts to seek out paintings, some of which he described in some detail, and to call on artists, including Britain's leading portrait-painter, Sir Joshua Reynolds.

Many of his observations, including some of the most detailed drawings in the journal, deal with technology and industry in a wide sense. Examples in Holland are the industrial mills at Zaandam and the locks and sluices in rivers and canals. In London, he comments on the steam engines pumping water from the river Thames, and on a new type of wheel for coaches made by the specialist firm of Jacob and Viney. At Woolwich, east of London, he describes in some detail how they bored cannons in the gun-foundry. With hindsight, this passage is somewhat macabre, as twenty years later the British navy would employ these high-precision cannons in its bombardment of Bugge's hometown, destroying his house.

Bugge himself never published his notes and probably only ever intended them as an *aide memoire* for his private use. On one occasion he drew on them to present an overview of English observatories that he had visited. This was in a Latin address to the university, which was published so obscurely that it has caused confusion among bibliographers.[48] When, two decades later, Bugge spent six months in Paris as the Danish representative in the international commission for the introduction of the metric system of weights and measures, he again kept a journal. This time, he prepared an edition, in the form of fictional letters, which appeared in Danish, German and English. From the latter, the historian of science Maurice Crosland published a selective edition because "no other contemporary account pays so much attention to the *scientific* life of Paris".[49] It will have become clear from this introduction, that much the same can be said of Bugge's journal of his study tour through Germany, Holland and England, and that his illustrated account can be read with pleasure and profit, both by specialist scholars and the general public.

47   McConnell 2007, p. 66–69, mainly based on Christensen 2001.

48   Bugge (1779b). The Copenhagen Universitetsprogram for 1779 contains on pp. 3–23 an address, delivered in Copenhagen on 10 May 1779, which is unsigned and untitled. From internal evidence it is evident that the speaker was Bugge, and the content matches the title *Descriptio historica observationum instrumentorumque astronomicorum, maxime Anglicorum* ('Historical description of astronomical observations and instruments, mainly in England') which we find in lists of Bugge's publications in Andersen 1968, pp. 82- 86, and Thykier 1990, pp. 447–451. We have been unable to locate what is found in these same lists as *Programma invitatiorum inaugurale 1779 continens Descriptionem itineris sui ad Anglos* ('Inaugural programme for 1779 containing a description of his journey to England'). We conclude that the two publications are in fact one and the same.

49   Crosland 1968. The quote is from p. 5. On pp. 215–6, Crosland gives bibliographical details of the Danish (1800), German (1801) and English (1801) editions.

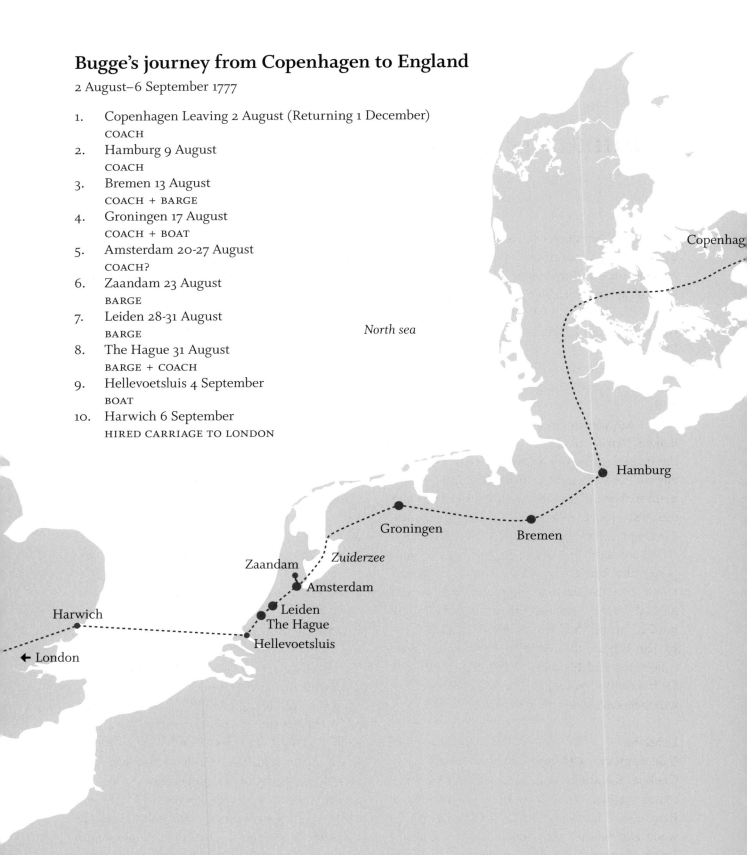

Bugge's journey from Copenhagen to England

2 August–6 September 1777

1. Copenhagen Leaving 2 August (Returning 1 December)
   COACH
2. Hamburg 9 August
   COACH
3. Bremen 13 August
   COACH + BARGE
4. Groningen 17 August
   COACH + BOAT
5. Amsterdam 20-27 August
   COACH?
6. Zaandam 23 August
   BARGE
7. Leiden 28-31 August
   BARGE
8. The Hague 31 August
   BARGE + COACH
9. Hellevoetsluis 4 September
   BOAT
10. Harwich 6 September
    HIRED CARRIAGE TO LONDON

North sea

Copenhag

Hamburg

Groningen

Bremen

Zaandam

Zuiderzee

Amsterdam

Leiden

The Hague

Hellevoetsluis

Harwich

← London

Summary of the Journal

Denmark, Germany, the Netherlands:
2 August – 4 September
Making observations about the towns and the state of agriculture, Bugge travels from Copenhagen via Hamburg and surroundings, Bremen, Oldenburg, Leer, Nieuweschans, Groningen to Lemmer. Here he crosses the Zuiderzee to Amsterdam, where he stays for nine days, including a day-trip to nearby Zaandam with its industrial mills.

In Amsterdam: the chart- and instrument makers Van Keulen, the menagerie of Blau Jan, the Amstel locks, the amateur astronomer Jacobus van de Wall and his observatory, the instrument makers Adam Steitz and Jan van Deijl and the instrument collections of the Mennonite Theological Seminary and of Ernestus Ebeling.

Three days in Leiden: Professor van Wijnperse, the University observatory and library, Lector Fas, the University botanical garden and natural history cabinet, the instrument maker Jan Paauw and the University cabinet of physics. In The Hague, the Stadholder's natural history collections and his palace Huis ten Bosch. Via Delft and Rotterdam, he reaches Hellevoetsluis where he embarks for Harwich.

London:
6 September – 5 October 1777
Chelsea: Royal Hospital, Ranelagh Gardens, the steam engines. Foundling Hospital. Pantheon; Benjamin Wilson's electrical experiments. Wedgwood and Bentley. Instrument makers Addison Smith, Benjamin Martin, Dollond, Nairne and Blunt, Samuel Whitford. Greenwich Hospital. Woolwich Royal Gun Foundry, Verbruggen's boring machine. Mathematician Samuel Dunn.

Electrical experiments with Nairne. Richmond: Demainbray and Rigaud; the King's Observatory and physical instruments (George III collection). Dollond, Haymarket. The Repository of the Royal Society of Arts. Watchmaker John Arnold. J.-H. Magellan. Glass and wheelwright's works at Blackfriars Bridge. Sculptor Charles Harris. Watchmaker Alexander Cumming and his observatory. Again John Arnold; pyrometer and pendulum experiments.

Portrait painter Sir Joshua Reynolds. Instrument maker Jeremiah Sisson, who has a model of Nathaniel Pigott's private observatory in Wales, and a transit instrument made for this. Instrument maker Jesse Ramsden. Royal Society Club, Mitre Tavern, Dr. Solander. Amateur astronomer Mr [William] Russell, shows instruments by Bird.

List of books bought in London, many from John Nourse and of instruments bought from the makers listed earlier.

Oxford, Cambridge, London, Greenwich:
5 October – 10 November
Eight days in Oxford: Radcliffe Library and Radcliffe Observatory. Detailed description, discussions with Prof. Thomas Hornsby. Bugge makes observations with the transit instrument here. Hornsby's apparatus for experimental physics. Sheldonian Theatre and Bodleian library, portraits; Colleges. Museum Ashmoleanum; the large mounted magnet. Further discussions with Hornsby on astronomical observations, including those by Roemer.

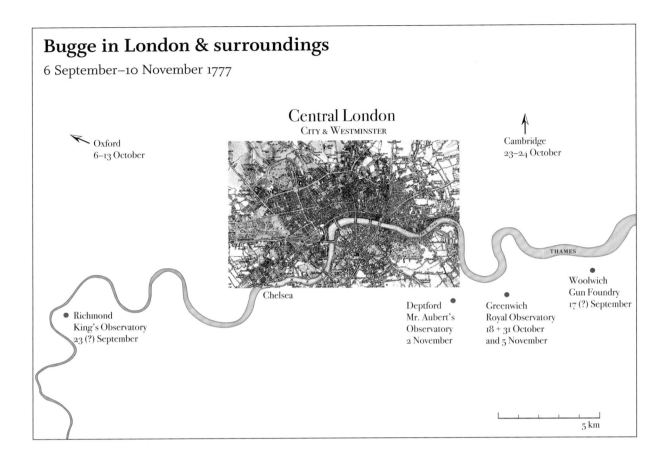

Bugge in London & surroundings

6 September–10 November 1777

Central London
CITY & WESTMINSTER

Oxford
6–13 October

Cambridge
23–24 October

THAMES

Chelsea

Woolwich
Gun Foundry
17 (?) September

Deptford
Mr. Aubert's
Observatory
2 November

Greenwich
Royal Observatory
18 + 31 October
and 5 November

Richmond
King's Observatory
23 (?) September

5 km

Return to London. Meeting at the Royal Society of Arts in the Adelphi. Instrument maker George Adams. Drawings of Greenwich Observatory. Mr Russell shows instruments and 'curious' books.

A few days in Cambridge. St. Johns College Observatory, where Ludlam made his observations. Christ College Observatory, Prof Shepherd is not there. The curriculum at the University.

Return to London. Astronomical Club. Lever's Museum in Leicester Fields. Another visit to Alexander Cumming to re-examine his barometric clock and Graham's clock with compensated pendulum.

Brief visit to Greenwich with Dr Maskelyne. Visit to Alexander Aubert's observatory, description of the instruments. Longer visit to Greenwich Observatory with Dr. Maskelyne. Inspection of the instruments made for him by Nairne and Blunt.

10 November departure from London. Travel via Osnabruck and Hamburg, arriving in Copenhagen on 1 December.

A note about money

The monetary standard in Denmark in Bugge's time was 1 rigsdaler (rdr) = 6 mark; 1 mark = 16 skilling; 1 skilling = 3 hvid.

On his journey through Germany, Bugge once gives a price in Mk L.: the Lübeck mark, double the value of the Danish mark.

In the Netherlands, Bugge paid in guilders, which were made up of 20 st(u)ivers. He sometimes indicated them as Gylden, sometimes as Fl. or Flor. In our translation we used the term florin throughout. Note, this is not to be confused with the Dutch silver coin florijn which equalled 28 stuivers.

In England, Bugge paid in pounds sterling = 20 shillings, 1 shilling = 12 pence (d). Often a sum is expressed in guineas, nominally 20 shillings but since 1717 circulating as legal tender at the rate of 21 shillings.

To compare prices in England and Denmark, equate 1 pound to 5 rigsdaler.

The Journal

[1 recto: some calculations; 1 verso, 2 recto and verso: blank]

Odense is poorly built, it has 50 master glovers, 50 married journeymen, 19 unmarried. A journeyman earns 3 marks a week. Three or four merchants trade in gloves on Hamburg and Lubeck.

Assens is a poor town. The entrance to Hadersleben might possibly be made navigable.

In the past, Appenrade had more ships than today; they sailed in cargo. *The country is very sandy. Stone farmhouses are covered with oak planks inside. The others are made of brick.*

From Aaresund to Appenrade most lands have been divided.[1] The king owns all the forests as far as beyond Schleswig; they are felled to produce ship timbers. The peasants are crown peasants all the way to Itzehoe.

Flensburg is an old town. There are fountains in the centre of the streets. The fields have recently been enclosed which gave rise to lawsuits.

Rendsburg and Schleswig are very beautiful towns; all the way to Schleswig the highway mostly passes through moors. None of the surrounding towns are walled, but the surrounding lands are good. Itzehoe seemed very poor to me. The Stör has been embanked. From there you pass through a fringe of

The journey to Hamburg
August 1777
From Copenhagen the 2nd.

---

1   The division of the common open fields around a village into farm plots began in the late eighteenth century, and Schleswig-Holstein was at the forefront of that development. As president of the Royal Agricultural Society, Bugge followed this with great interest.

August 1777
Hamburg

Hamburg at Der Schwarze
Adler

the Cremper Marsh. And immediately afterwards there are meadows, excellently laid out and well used, right to Hamburg. In between there are some moors and marshes which from time to time are cultivated.

Except for a few new houses, the town consists of old buildings. There is great activity, as one finds in a commercial town. The streets are mostly curved and narrow. The town has a garrison of 1200 men. At the moment it has no commander, but only a colonel; the police is excellent; public gardens are not allowed except near the castle of Hamburg. The town gates close at 8.30; besides the patrolling watchman and the night watchman, many other watchmen are spread over the town.

The outward appearance of the *Town hall* is rather beautiful and new, but the inside is gothic. There are two council halls, one for the sage council, consisting of 4 mayors and 24 councillors and the Notarius; the other hall is the meeting place for the citizens.

The church stands close to the town hall

and is an old building decorated with statues.

The *Baumhaus* is a very large building standing at a branch of the Elbe. The house belongs to the town, but it has been let on lease. Actually it is a public house, divided into large halls for assemblies, balls, etc. Upstairs there are balconies, and at the very top a dome. Everywhere the view is fine.

The *Commercial School* is a private institution led by professor Bühde and doctor Ebeling.[2] The first was in Holland, but I spoke with the latter. He showed me syllabuses adapted for commercial purposes. Further he showed me the whole house. The fees are 1000 to 1500 mk.L *333 Rdr. 500 Rdr.* per pupil. Registers are kept of the pupils' diligence and behaviour, and these are sent to the parents every three months. Those who are not diligent or have been naughty are sent away again. This was, for instance, the case with Brown's son from Copenhagen.

The *Bourse* is a rather large building erected above the water. It has a fine and open colonnade.

---

2   Christoph Daniel Ebeling (1741–1817). Expert on geographical studies, especially America. Held for 33 years the chair of history and Greek in the Hamburg gymnasium. Also superintendent of the Hamburg library.

Hamburg
August 1777

A *water mill* stands immediately at Der Schwarze Adler. It is a very large, brick building with 6-8 wheels. The promenade is called the *Jungfernstieg*; a rather long, but narrow and uneven avenue along the Alster; it is also allowed to promenade on the ramparts of the town. The number of people walking back through the gates on Sunday evenings is amazing.

The playhouse is situated in a miserable corner and has only one exit. The indoor decorations are tasteless. This evening's performance was *Zu gut ist zu gut*, translated from English. The director Schröder[3] played very well. And a certain Brukman played the part of Lofstead Bille fairly well. The females were rather poor. The ballet was Vauxhall; the representation was good; but the composition and the dancing were poor. The best dancers were the director's wife and a Frenchman.

On the other side of Hamburg, the merchants have several gardens on which they spend a good deal of money.

---

3   Friedrich Ludwig Schröder (1744–1816), celebrated actor and theatre
     director, who introduced Shakespeare to the German stage.

Through Altona (where there are many beautiful buildings and gardens, and an amazing number of Jews) and the nearby village Ottensee I went 1 mile to Nienstad. The road runs along the Elbe river. Here as well as at Nienstad the view is remarkably beautiful all the way to Buxtehude and Stade. In this latter village there are several beautiful gardens and summerhouses.

Wandsbek[4] is a beautiful village owned by G.H.R. Schimmelman.[5] Here he has recently built a beautiful mansion in a large garden with many beautiful and regular paths, statues, fountains and pleasure houses. Part of the garden belongs to a small forest. Everything is very tastefully arranged. Nearby there is a *creamery*, which gives a good yield. The village has straight and broad streets and beautiful houses, for whose erection the Privy Councillor has paid an advance. In the village are a lot of artisans and artists who are smuggling their goods into Hamburg.

Outside the Stone Gate (Stein Thor)

1777
Hamburg

---

4   Bugge wrote Vanstæd, he may have heard it wrong.

5   Carl Heinrich von Schimmelmann (1724–1782), since 1762 'Freiherr', since 1765 'Geheimrat' (the abbreviation before his name), since 1779 'Graf' (count), was a successful German entrepreneur. As Danish Treasurer he negotiated the so-called Gottorp Settlement (1762) between the city of Hamburg and the Danish government. That same year he acquired the estate of Ahrensburg and nearby village of Wandsbek. He modernized the mansion and stimulated small industries on his grounds. Klessmann 1981, pp. 546–549, with reproduction of his portrait in oil by J.G. Ziesenis.

the inhabitants of Hamburg have several beautiful little gardens on which they are spending enormous amounts; the extent of the village and of the fortifications is very large. In the evenings the citizens are occupying the ramparts. Everywhere in Hamburg luxuriousness of clothing, eating, and game-playing have increased enormously. Moreover, there is little politeness and kindness to foreigners, and these are constantly pestered by hucksters, tailors, and Jews, all of them waiting to get profit.

I met the country syndic Jacobi from Zelle,[6] the secretary Walter from Copenhagen,[7] and the Countess Lerche. The first mentioned had much pleasure during Walter's absence.

The last evening I went to the Vauxhall. It is a small garden which is fairly well illuminated. There was music and singing. The audience was not numerous, but beautiful. The entertainment was poor. This evening some young Englishmen from the Commercial School behaved rather badly, pushing other people etc.

6   Andreas Ludolf Jacobi (1746–1825), a lawyer who in 1768 was sworn in as solicitor at Celle, a town located in the southernmost part of the Lüneburg Heath. In 1775, he became "Syndicus der Lüneburgischen Ritter- und Landschaft", which he remained for fifty years.

7   Thomas Christian Walter (1749–1788), musician and musical director of the Royal Theatre in Copenhagen until 1775, when he left his wife, the celebrated actress Caroline Walter (1756–1826), to travel abroad.

The 13th I went to Bremen by ordinary stage coach. For the trip you have to pay 3 rigsdaler in advance. First you leave at 12 o'clock for Harburg (two times a week, on Saturday and Wednesday). This short voyage by boat is not uncomfortable because of the many branches of the Elbe, which form islands that are inhabited. On one of these *which has been embanked* the inhabitants of Hamburg have their timber stores as well as a good saw mill. There was a fair wind, and the crossing took about 2 hours.

Harburg is a small fortress, but the town itself is not at all extraordinary.

From Harburg to the first station, Dobstæd, a peasant town, is two miles. The country is hilly, sandy, and covered with heather.

To the next station, Rothenburg, a small town, 3 miles. To the third station, Ottersberg, 2 miles. The fourth station is Ottersberg [sic] 3 miles, and then you are in Bremen.

All the way through the Hanoverian region the country is as described in the following:

It is covered with moors, which belong to the peasants, but others have the right to keep sheep on them. This makes the abolition of the open field system almost impossible. The sheep are poor and small; and there are more black and spotted than white ones. There are almost no enclosures except for the gardens of the farms. The style of building is much like that in Holstein. There was nothing but rye, buckwheat and oats, and there are very few meadows and marshes. Once you get up to the high land, it is mostly very even. There is no concentration of forests, but small separate groups, mostly beeches and a few oaks. The latter are planted and are taking fairly well. Everywhere they get only very small and are left for too long.

Toward Bremen the country is better and consists of marshy soil; everything is enclosed and planted.

The 14th I arrived at Bremen at 4 o'clock in the afternoon. The town is rather beautiful. The houses have been very nicely painted outside as well as inside; the streets are broad. The interior of the churches is not only gothic, but also very poor as regards the pews, etc. The *König von Preussen* is a good inn. From here there is an ordinary stage coach to Leer in East Frisia. The journey costs about 3 rigsdaler in total.

The 15th. At 8 o'clock in the morning I left Bremen. The stages are: 2 ½ miles to Falkenburg, 2 ½ miles to Oldenburg, 3 miles to Marburg,[8] 3 miles to Leer, that is 11 miles in total.

One mile from Bremen is Delmenhorst, a poor town, but the surrounding land is good and well cultivated. The peasants do not perform statute labour, but pay fixed dues to the squire.

Further away into the Oldenburg region, the country gets sandier, now and then there are moors, which could

1777, August
Bremen

---

8   The road went through Harkebrügge, which Bugge must have misheard as there is no Marburg in this region.

August 1777

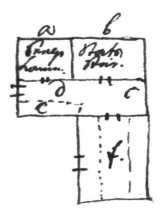

be cultivated.

The farmsteads in the Oldenburg and Delmenhorst regions are of the following construction: a. the peasant's bedroom; b. the best parlour; c. the dining and living room; e. the cattle house; f. threshing floor, barn, and cattle house. Chimneys are not yet in use.

Oak trees are planted around almost every farmstead, and they grow well; they even sell oak timber to Holland.

Oldenburg is a beautiful small town, and the castle is now being enlarged. In the past, the town was fortified by five double ramparts, but now it has only one single rampart. All corpses are buried outside the town, where a nice small church has been built for funeral sermons. At the vicarage is a lime tree, whose lower branches have been arranged as a gallery with a balustrade; the road to the churchyard goes through a very nice arcade.

In spite of the fact that you usually get good horses at the stages, and that 5 horses are hitched to those big freight carriages, you never travel faster than at a walking pace all the way from Hamburg to East Frisia or Prussia. The coaching regulations are bad, and the postmasters say that they are not obliged to show them. The accompanying Schirrmeister[9] does not care for anything but the mail-box. The mail coach drivers are rough, and there is nothing one can do but let them drive as they like.

In East Frisia the coaching service is well-organized, and there you travel very well. The highways are bordered with trees, but not

---

9   The German title for a minor officer responsible for vehicles.

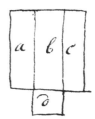

with ditches. The trees are crooked and badly placed, and most of them have either died or have been run over.

At first the country is very sandy and heathery, but towards Leer it gets better. The layout of the East Frisian farmsteads is as follows.

a. barn
b. threshing floor
c. cattle house
d. farmhouse

Here chimneys are in use.

At Kloster Barthe[10] a home farm has been established on behalf of the king. The annual rent is 800 rigsdaler.
The building is as follows:

a. farmhouse
b.b. threshing floor and fodder bay
c.c. barns
d.d. cattle houses

Elevation

Ground plan

10   Bugge seems to have written 'Back', but a web search suggests that this was Kloster Barthe near Hesel, a medieval monastery turned into an agricultural enterprise with the secularization during the Reformation.

Leer is a nice small town, which conducts an important trade with Holland in linens which are mainly woven in this town. Part of the yarn is spun in the town, but most of it is spun in the country by peasant girls and boys. Here the Dutch neatness can already be observed.

From Leer to Nieuweschans is two miles. By the ordinary stage coach the fee is 3 marks. *From Leer the East Frisian marshland begins. A large part of the road goes over an old dike, which is said to have been constructed in 1530. Outside it are two more dikes, and the new farmsteads are placed at equal distances from each other. In this part of East Frisia it is still possible to reclaim considerable amounts of land from the sea.*

I arrived at Nieuweschans the 16th in the evening; and spent the night there. The fortification and the town are both small. The canal barge left Nieuweschans the 17th at 7 o'clock in the morning and went to Winschoten. The voyage costs 18 stivers. In this town there are considerable tileries.

From Winschoten to Groningen you have to pay 18 stivers for the canal barge. The voyage is extremely agreeable, as the canal is in most places planted with trees and has a continuous variation of buildings, mills, bridges and small locks.

On the 17th at 4 o'clock in the afternoon I arrived at Groningen. The town is very nice *and well fortified*. The trees placed along the canals are kept regularly pruned, which you do not even see in Amsterdam. The town hall had been pulled down, and some booths have been erected in the square. It was said that they were short of money for the building.

The Academy had about 150 students. Time did not allow me to visit the library.

Groningengetslargeprofitsfromitshosieriesandweavingmills. *The neighbourhood of Groningen is very nice. 1/8 mile outside the town is a small wood, Sterrenbos, decoratively intersected by avenues; this is the public promenade of the town.*

On the 18th I went by special stage coach (a post barouche with 4 horses) from Groningen to Lemmer and had to pay 18 florins for about 11 Danish miles. The road passes through the provinces Groningen and West Frisia; especially in the first and in the first part of the latter province there are many large moors; but otherwise there are many beautiful enclosures, quickset hedges, and good agriculture.

Near their houses the farmers plant large numbers of oak trees which are growing very badly. The highways have been planted in the same way.

In West Frisia the road goes through marshland which is so low that it can not be used as arable land; but on the other hand, there is an abundance of peat soil, and everywhere one saw large peat stores, built of wood in such a way that the air could pass freely through them. They have been built at the canals in order to facilitate transport across the Zuiderzee to Amsterdam.

On the 18th in the evening I arrived at Lemmer where I at once boarded the ferry which leaves for Amsterdam every night. The voyage across the Zuiderzee lasted until the 20th at 5 o'clock in the morning, after a voyage of 11 Danish miles, partly with calm, partly with contrary wind.

August 1777

August 1777

Amsterdam
20th August 1777

▲ *Octant signed 'J.V.K.1762', length of*
alidade 574mm. Museum Elburg.
Fifteen years after the firm of Johannes
van Keulen made this octant, Bugge
visited their Amsterdam shop and
inspected the octants on sale. Octants
served to measure the angle between
a celestial body and the horizon, from
which the observer could calculate
his latitude. Also known as Hadley's
quadrant, after the man who invented
it in 1731, it was a major improvement
in navigational instrumentation.

The fare is 1 ducat; a berth in the deckhouse costs 28 stivers; and in the steerage 1 florin or 20 stivers; I chose the latter place and found very good berths there; besides I had to pay 6 stivers for my suitcase and 9 stivers transit fee.

In Amsterdam I lodged at de *middle bible*, which is a very nice inn.[11] At the booksellers Harrevelt[12] and Changuion,[13] I got their catalogues only with difficulty. At the first mentioned they sell Le Guide d'Amsterdam, 1772, which is a very useful book for foreigners.

At *Covens and Mortier*,[14] I got seven special maps of the United Provinces and a general map, as well as a special map of Amsterdam.

At Johannes van Keulen en Zoonen,[15] I bought some charts and sailors' books. I also saw several *Hadley's*[16] octants *of ebony with brass edge* of his own construction. They are rather good. On most of them the eyepieces were too large. The smallest ones, which have a radius of 7 inches, cost 18 florins.[17]

11　The inn De Middelste Bijbel in the Warmoesstraat had been established in 1646. Wijnman 1971, p. 115.

12　E. van Harrevelt, in business 1747–1779 in the Kalverstraat, opposite the Gaperssteeg. Brink and Werner 1989, p. 109.

13　Daniel Jean Changuion, w. 1766–1803 in the Kalverstraat. Brink and Werner 1989, p. 106.

14　The firm founded by Johannes Covens (1697–1774) and Cornelis Mortier (1699–1783) was the foremost map publisher in 18th-century Amsterdam and was to remain in business until 1866. Brink and Werner 1989, pp. 43–44.

15　The chart-makers firm of van Keulen was established in a building named 'De Gekroonde Lootsman' (The Crowned Pilot). In the year of Bugge's visit, the firm printed a trade catalogue in which it called itself 'Boeck- en Zeekaart-Verkoopers, Compasse, Octante- Graadbooge en Mathematische Instrumente-Maakers' (Book and Chart-Sellers, and Makers of Compasses, Octants, Cross-staffs and Mathematical Instruments). van Keulen, Mörzer Bruyns, Spits 1989.

16　Bugge wrote "Halley's", probably a slip of the pen, as it is unlikely that he confused Edmund Halley, of comet fame, with John Hadley, the inventor of the octant. Later in London, he was to buy a copy of the treatise on Hadley's octant by the Cambridge mathematical practitioner William Ludlam.

17　For the use and production of octants in the Netherlands, see Mörzer Bruyns 2003, which mentions Bugge's visit to the van Keulen shop on p. 121 and reproduces the list of octants in the firm's 1777 catalogue on p. 133.

The pavement is excellent and consists of small, oblong squared stones, placed according to the same rules as in France. There is a footpath or trottoir *of 4 feet* [18] on both sides; and in the broad streets there is a third one in the middle.

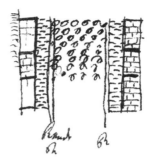

Every morning a man arrives with a closed carriage with several compartments. He removes the refuse from the houses and collects sweepings, ashes, etc. However, you often find shards and other impurities by the side of the streets. The gutters are almost edge gutters and are rather unclean. Most of the houses probably have outlets to the canals. All kinds of impurities from the houses are thrown into the canals; whereby unhealthiness and stench are much increased.

I saw:

The Bourse, the whole square was crowded with people.

Oude Mannen en Vrouwen Huis [Old Men's and Women's Home].

Raad or Stad Huis [Town Hall].

Spin Huis [Spinning House]

The big Wees Huis (Maison des enfans trouvés) [Orphanage]. 1200 to 1400 children stay there until their 22nd year.

Hortus Medicus: many rare plants and trees. Coffee. Tea.

Blau Jan: a small menagerie of animals;[19] the most distinguished ones were a lion and an American

---

18 Bugge writes '2 alen'. One 'alen' is a Danish measure of length equalling two feet.

19 The menagerie Blau(w) Jan was located in an inn 'De Hoop' along the Kloveniersburgwal and was a popular attraction from its foundation in 1698 until it ceased to operate in 1784. Engel 1986, pp. 30–31.

bear; a [illegible word]; many sorts of monkeys; parrot, cocka-toos, etc. There was also a dwarf, 22 inches high; his head and chest were of natural size; the lower part of the body was short and badly proportioned. He was married to a girl from West Frisia; and by her he had well-formed children.

The Oude Kerk [Old Church] is beautiful. In two of the large church windows several historical scenes have been burnt into the glass; especially the drawings and the colours are beautiful.

The *Town Hall* is a noble building. The two large salons on the second floor are covered with cornices, statues, bas-reliefs, marble altogether. Everything is in good taste.

In the rooms there are very beautiful paintings; but to me they seemed to have been covered by too strong a varnish.

The Portuguese synagogue is very large; and the interior is simple and beautiful. The German synagogue consists of two small ones. The building and the interior are not so beautiful, but the singing and the service are much more beautiful here than at the Portuguese one.

The most common locks in

West Frisia, Groningen, Holland, and Amsterdam are as shown in the following ground plan.

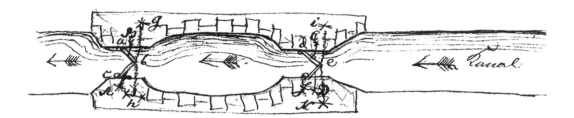

ab and bc, as well as de and ef, are two lock gates which are pressed together by the current or the course of the river. g, h, i, and k are winches for lifting the gates.

As regards the profile of the lock-gate, note that on the large lock-gates ab, cb, de, and ef there is a small opening or gate below the water. These are necessary in order to obtain the same water level inside and outside the lock during the passage of ships. They are opened with the winches A, B, C, D either by means of gears, or by some other mechanical construction.

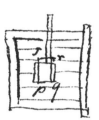

I have often seen that the water level in the canal outside was 2, 3, or 4 feet higher than inside the lock eb itself; and, necessarily, higher inside the lock than at the left side of e. You

August 1777
Amsterdam

often find locks with two gates, and furthermore they are strengthened by girders. In the summer the central gate is always open.

Many locks are of the same width throughout; everything is made of brick; the edges are covered with sandstone; on the sides there are holes with iron bars or gratings, in order that the walls shall not be damaged during the passage of ships. Above there are iron rails, so nothing is lacking that may be necessary, useful or neat.

Zaandam

On the 23rd I went to Zaandam[20]; on the map it is called Saenredam, after the bay or broad stream de Saen or Zaen. It is divided into East and West Zaen. This village is particularly lovely; the streets are covered with bricks all over, yellow on both sides and red in the middle. It is intersected by innumerable small canals, planted with trimmed trees.

20 Bugge spelled this place-name as 'Serdam' throughout.

The houses are small, but neat. They all have small, nice gardens or parks, usually facing the street and decorated to everybody's different taste. In the most beautiful ones the ground has been covered with grey sand or seashells; the ornaments are surrounded by box trees, and inside they are covered with white stones and pit coal, placed in neat rows. You cannot imagine a lovelier sight than the many houses, gardens and summerhouses on both sides of the Zaen, and the constant alternation of meadows, canals, mills, ships, etc.

At Zaandam there is an almost unbelievable number of windmills, saw, flour, oil, tobacco, chocolate, tanning mills, etc.[21] I saw a few sawmills with nothing particular about them. A 50 feet long oak trunk was cut into four planks in two hours by means of five saws. By the same wind went another saw frame, whereby large ship timbers are sawn through once. Such a

21 On the industrial mills active in the region around the river Zaan in the 17th–20th centuries, see Boorsma 1968, with technical details on the operations inside the types of mill that Bugge describes, viz. sawmills (pp. 34–38), oil mills (pp. 42–47), paper mills (pp. 47–53) and tobacco or snuff mills (pp. 60–62). For paper mills, see also Voorn 1960. Each mill had its own name, but as Bugge mentions neither these, nor the names of millers or proprietors, it would be hard to identify which specific mills he inspected.

Zaandam

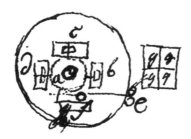

mill costs 12000 florins and makes an annual profit of 2 to 3000 florins.

A *Paper Mill* was of the following construction: a. is the shaft with a big wheel, which drives two Hollanders or cylinders with knives, b, c, and d. The rags are put into b, and, depending on the circumstances, the pulp is worked in b alone, or in d as well. e. are two pumps outside the mill which are also driven by the wind, and they pump up the salt water into the filter boxes g.g.g. where it is filtered several times through sand boxes. From there it runs to a reception box below the pump f inside the mill where, by the proper motion of the machine, it is pumped into the reservoir h *where it runs through many hand cutters* from where it is distributed to the Hollanders b, c, and d through pipes and taps. In spite of the very brackish water the paper becomes excellent.

In the Tobacco Mills which I visited they make Spanish tobacco, [two illegible words] and Rappe.

1. *Spanish tobacco.* It is mixed with 1/3 of the moss or fungus that grows on

oak trees. First it is macerated in water, and then dried like the tobacco, on the open space abc which is covered with Dutch tiles. Then the stems are removed by means of a small closing mechanism in R. ef and gh are two perpendicular millstones which, by means of two circular scrapers, put the tobacco back under the stone. From there it is carried to the sieve mn, and is first thrown into the sloping funnel which is continuously shaken by the skew gear m. The sieve is very fine. Various sorts are produced: 1. coloured with red, 2. coloured with yellow, 3. completely natural coloured.

2. Ordinary tobacco or Dutch Rappe. First it is stamped or cut by means of stampers with sharp spades at the end in lm; as soon as the tobacco has a certain fineness it falls through, and from there it is carried to the sieve pq which is not as fine as that used for the Spanish tobacco; the fine and finished tobacco remains in the sieve box, whereas the coarse tobacco, mainly pieces of stems, falls out at the end g. It is carried under the stones fc and hg in order to be ground. As it is thereby packed and pressed together in small hard lumps,

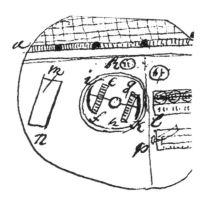

August 1777
Zaandam

it has to be put under the knives or stampers lm again, from where it is again put into the sieve.

One of the best *oil mills* has a double work of the following construction:

The second half of the work.

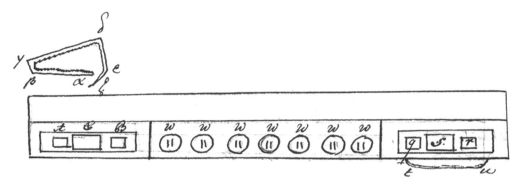

The first half of the work.

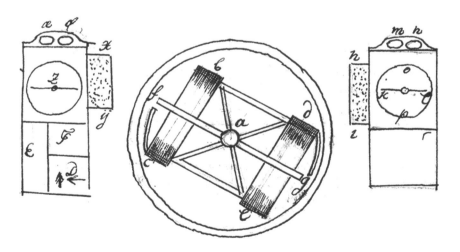

The shaft a drives two very big perpendicular millstones cb and de fitted with two collectors or scrapers cf and dg. Here the rape-seed is crushed first of all. From here it is put into the box hi, and a certain quantity is filled into an iron saucepan, placed on a fire which is also heating the box hi. For without heat the oil cannot be extracted. The rapeseed in the saucepan is stirred by means of the horizontally rotating stick Kl which is driven by the mill itself. m and n are two holes, below which two bags are hanging, each of them containing the same quantity as the above mentioned box hi. When the bags are full, each one is put into a *pyramidal* leather case αβγδεζ, lined inside with a sharp and rough horsehair tissue. The bags with the rapeseed are put into the holes q and r, and a slowly falling stamper with a conical projection S is pressed down between them. Then the linseed oil runs down into the vessel Eu placed below.

The chief assistant in the mill rings a bell, when the cakes have received a certain number of strokes from the stampers.

When the cake is taken out, it is hard and about ½ inch thick, as compared with nearly 6 inches before the stamping; these cakes still contain much oil. Therefore, they are placed under the spade or knife stampers w.w.w. After being

August 1777
Zaandam

August 1777
Zaandam

Amsterdam

pulverized here they are carried to xy, and are measured out in the iron saucepan Z, heated and stirred and again filled into the bags and the leather case; they are placed in the boxes A and B and are again squeezed by the conical wedge, when it is pressed down by the stamper. When they are removed, the lower pointed end still contains oil; this part is cut away by means of a perpendicularly placed knife D. The oily pieces are put into the box F; from where they are later placed below the stamping knives, heated and squeezed for the last time. The useless parts of the cake are put away, and used as cattle feed.

The old playhouse in Amsterdam burned down two years ago,[22] because the chandeliers had been hoisted up too close to the ceiling. There was only one exit, and many people were burnt or injured.

The new playhouse has been erected at the Leiden gate and is built of wood. The interior decoration is beautiful; there are two tiers of boxes; painted … and decorated [only partially legible] with gilt

---

22   This is not quite correct; it had burned down in 1772. Ironically, the new theatre that Bugge saw, built in 1774, was also to burn down in 1892.

festoons. The ceiling is of plaster. In front of the theatre there are two columns of porphyry with marble capitals, between them are two beautiful statues, but on the whole the proportion is not as beautiful as at the Danish Theatre; the room might well have been fitted with 4 tiers of boxes (inclusive of pit boxes).

On the night when I saw the playhouse they played Conradus, a tragedy translated from the German, followed by a Dutch play, the Ridiculous Damsels,[23] and finally a ballet. All this lasted from 5 to 9.30. The dresses, music, scenery paintings and dancing were very beautiful; but the action was vulgar, and the declamation poor.

A French merchant, Peaumé, who stayed with me at the Middelste Bijbel, introduced me to a rich merchant, van de Wall, who lives immediately outside the Leiden gate; he is not only a lover of the sciences, but constructs mathematical instruments himself.[24] In his garden he has built a small observatory, whose roof can rotate over rollers by means of pulleys. The opening in the meridian or Crenae throughout is plane copper plates fitted into grooves, and these copper plates can be hoisted up

August 1777
Amsterdam

---

23 Possibly *De Belachchelyke Hoofsche Juffers*, a farce by Pieter de la Croix after Molière's *Les précieuses ridicules* of 1659. The Dutch edition was published in 1685, with a reprint in 1753. Personal communication A.J.E. Harmsen, Leiden University.

24 On the Amsterdam merchant and amateur astronomer Jacobus van de Wall (1700–1782), his observatory and his large reflecting telescope, which he bequeathed to Leiden University, see Zuidervaart 2003, 2004 and 2007. It is now in the Museum Boerhaave, Leiden.

August 1777
Amsterdam

by means of pulleys. A specially comfortable appliance has been constructed to sit in or lean against when observing in different heights. On the inner sides of two S-shaped boards AB and CD, spaced about 3 feet apart, parallel lists have been fixed two by two at intervals of about 5 inches. If boards are placed between these lists, for example in BD, EF, or GH, you can easily make observations, either sitting or standing.

The most remarkable object at this dignified man's house is an 8 foot reflecting telescope constructed by him. It works very well and is of excellent workmanship.

I will begin by describing the support. Its lower part consists of 4 iron feet DC, CE, FC and CG, fitted with adjusting screws D, E, F, and G; these feet are jointed at C which is a fixed cylinder; branching off from each of these four feet are horizontal iron bars, joined at H which is placed immediately below the centre of the hollow cylinder C. To these four feet is

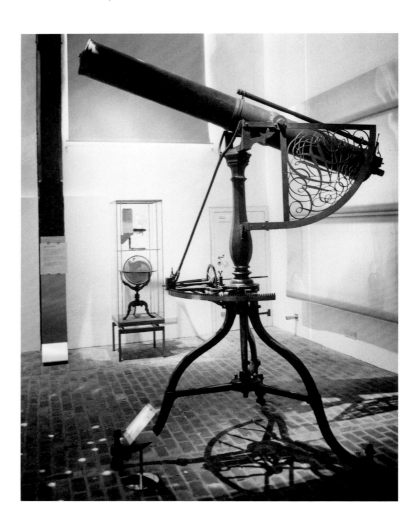

Eight-foot reflecting telescope, made by Jacobus van de Wall in c. 1742, length of brass tube 215cm. The quadrant (radius 52cm) allows measurement to 5 arc seconds. Museum Boerhaave, Leiden. In Amsterdam, Bugge visited the private observatory of the aged amateur astronomer Jacobus van de Wall. The main instrument was this enormous telescope which his host had made himself, and which Bugge considered to be 'of excellent workmanship'. After Van de Wall's death in 1782 it was transferred to Leiden. ▶

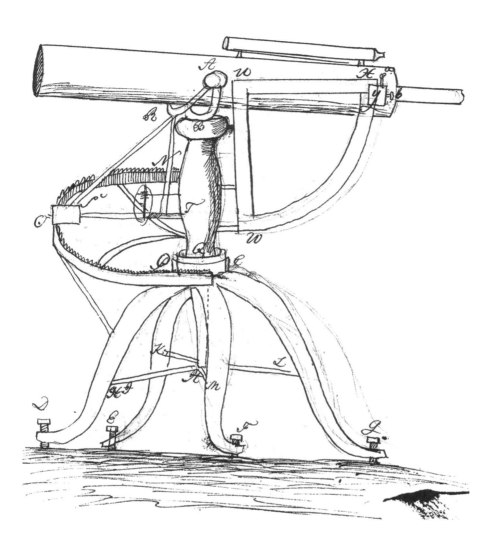

also fixed a horizontal half-circle NOP, whose upper edge is toothed and graduated. Nothing more is attached to the feet.

On the cylinder of the telescope two pivots have been fixed to the axis A. These are placed in the fork AB which is fixed to the vertical axle BQH, moving in the hollow cylinder QC and in the hole H.

Branching off from A is another fork RS, supporting a gear box S which is furthermore connected with the disk ST.

The box S contains a wheelwork which can be pressed down into the teeth of the half-circle NOP and be moved by a fine screw. It also has a pointer which marks the degrees on the edge NOP, or indicates the azimuth once the telescope has been placed in the meridian.

Furthermore, a quadrant UWX, divided in 90 degrees, has been fixed to the vertical axle BTQH. The edge WX is toothed and meshes into an endless screw a placed in the box Y. By means of a screw b these teeth can be made to mesh or not. It is easy to see that the telescope can be carried up and down by this quadrant, and the altitude can be measured by means of the quadrant.

In order to make this instrument more stable, a wall has been erected which has no connection with the wooden building; on the wall this heavy machine is placed.

Mr van de Wall maintained to have observed that the rings of Saturn get thinner towards the edges and get thicker towards Saturn.

And that Saturn was full of circles parallel to themselves and to the ring.

Dimensions *Amsterdam measures*:
The large mirror.
 Focus = 8 foot.
 Diameter of the aperture = 9 inches
 Diameter of the hole = 2 5/8.
 Focus of the small mirror = 8 inches concave
 Magnification = 300
 The oculars.

August 1777
Amsterdam

The telescope is Gregorian

August 1777
Amsterdam

Mr van de Wall had also several smaller telescopes made by himself. He told me that when he heard that Short[25] had constructed a 12-foot telescope, he himself had been constructing one of 18 feet. Its mirror was already completed. But as the other parts of the telescope and especially the support would be too heavy and strong, he had given up the work.

It must also be noted that with the great telescope Mr van de Wall is able to magnify up to 600 times; but then he changes it into a Cassegrainian telescope[26] by changing the focus of the small convex mirror from 6 to 4 inches. Furthermore, the air must be extremely pure. *NB. to see the object in upright position by means of two oculars.*

Mr van de Wall, who is now in his seventy-seventh year, showed me an excellent piece of flint glass from England. It was 9 inches in diameter *and 2 inches thick*, and it was completely free of veins and striæ. Compared with this he showed me a piece of crown glass, full of veins, which he had ordered from Germany. If it had been good, he would have made a Dollond telescope[27] with it.

[19 recto] contains two standard geometrical formulas
[19 verso] contains various pencil sketches which cannot be readily identified.

25   The famous manufacturer of reflecting telescopes James Short (1710–1768) constructed over 1300 instruments.

26   A type of reflecting telescope as invented by a Mr. Cassegrain around 1670. It differed from the Gregorian telescope as designed by James Gregory some years earlier in that it used a convex mirror for the secondary ellipsoid rather than a concave one.

27   A refractor with achromatic lens-system as patented in 1758 by John Dollond (1706-1761) and his son Peter Dollond (1731–1820). The patent was both infringed upon and much resented by other opticians until it expired in 1772.

[20 recto] some notes on expenses

[20 verso] blank

▲ *Magnetic toys by George Adams, London, 1765, length of boxes 350mm. Science Museum, London. They are similar to what Bugge saw in Steitz' workshop. The demonstrator could determine the position of the pieces as if he were able to see them through the closed lid.*

I visited Mr Adam Steitz,[28] an instrument maker living in Runstraat; this man is a good mechanical genius, and his products are of perfectly good workmanship, but they seem very expensive to me. For a balance illustrating the theory of the lever he charged 80 florins. For a model of a fire-pump, about 1 foot long and 1/2 foot high, he charged 160 florins.

He showed me a nice little trick. In a box there were four pieces with numbers on them, 1, 2, 3, 4. You were allowed to place them in arbitrary order, and then hand the box back to him closed. Then he looked through a brass tube CD which was seemingly dark, and was able to tell which numbers were in the box. The whole trick is that the pieces 1 and 2 are fitted with artificial magnets whose north poles turn in a certain direction, in No. 1 downwards, in No. 2 to the right, etc. When the screw gh is turned, light will fall on a small compass af, from whose direction

Amsterdam 1777

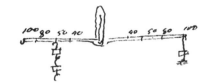

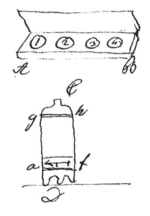

[Note: the diagram in the bottom corner is unrelated to the text; it may show some device to demonstrate the power of water or steam, seen from above; the central caption is 'water kettle', in the caption at the top the first word is also 'water'.]

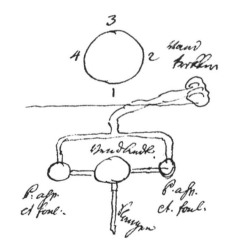

28 On this Amsterdam instrument maker, see Rooseboom 1950, p. 126. Surviving products include a solar microscope in the Museum Boerhaave, Leiden, and a centrifugal machine, see Bos 1968, p. 56. Several of his demonstration instruments, some made together with his fellow townsman J. Kampman, are in the Teyler's Museum, Haarlem; they were bought by the scientist Martinus van Marum at the Ebeling auction sale in 1791 (on which see below), see Turner 1973. Steitz also worked for Jacobus van de Wall and in 1782 transported his large reflector and observatory building to Leiden, see Zuidervaart 2004, pp. 436 and 443.

the position of the north pole is determined, and thus it is possible to say whether it is 1, 2, 3, or 4. [29]

He also showed me a small machine for ejecting a glowing parabola. The sphere cb of about 2 inches has a tube at its end which is closed in a and is 4 inches long; at d is a pin with a very fine hole which can make 30, 40, 60° at ab depending on whether you want the parabola to be long or short; when the sphere bc is heated, and ad is held in cold spiritus vini, the sphere is eventually filled, except that some air remains in it. Then the bowl mno is filled with spiritus vini which – when ignited – through its heat expands the air in the sphere at c; and through the elastic pressure spiritus vini spouts out in a parabolic curve.

Mr Steitz arranged for me to see a collection of physical instruments which have been acquired with a contribution of the Mennonites.[30] It is a kind of gymnasium for Mennonite students where public lectures on science are held.

Mechanical cart by Kampman and Steitz, Amsterdam, c. 1750-1775, height 180mm. Teyler's Museum, Haarlem. Bugge saw such a device demonstrated in the workshop of Adam Steitz. The cart is set in motion on a table and a mechanism shoots a ball upwards, which will land back in the cup. The uniform horizontal motion of the cart and the uniformly accelerated motion of a vertically projected body cause the ball to describe a parabolic trajectory. ▶

29   For more details on the magnetic toys made by George Adams, illustrated on page 36, see Morton and Wess 1993, pp. 444–445.

30   From 1761 onward, the Mennonite Theological Seminary built up a collection of philosophical apparatus, which was the first institutional cabinet of this kind in Amsterdam. Steitz was principal supplier and also maintained the collection, which was sold by auction in 1828. See Zuidervaart 2006.

This collection is very elegant and rather complete. Most of the devices are like those described by Mr 's Gravesande[31] and they have been constructed and improved by Mr Steitz. For instance, he has given the barrels of the air-pump a diameter of 4 inches, which facilitates the opening and closing of the valves. He has also very much improved 's Gravesande's machine for centrifugal forces. He has also constructed a mechanical cart, which shoots a ball a vertically upward *a spiral [and a] bar of ivory* and demonstrates that it falls down again at or on the cart.[32]

I also visited Mr Jan van Deijl and Son in the Vinkenstraat at the Haarlemmer Dijk.[33] They both have a good theoretical background and make good Dollond telescopes. The son claims to be able, by means of simple geometrical construction *and simple algebra*, at once to find out the curved surfaces, the focuses, etc. of both glasses. He also claimed to have found out how to construct a telescope showing the objects in upright position, by means of 4 ocular glasses.

31   Willem Jacob 's Gravesande (1688–1742), professor of mathematics, astronomy and natural philosophy at Leiden University. He devised a large variety of demonstration instruments for physics experiments, which he described and illustrated in his text-books.

32   Bugge wrote *en Spiral bom af Elfenben* in the margin, which means *a spiral bar of ivory*, perhaps to remind himself that it was fitted with a spring and an ivory fusee. The cart illustrated opposite is decribed in Turner 1973, p. 155. For other examples see ibid, Bos 1968, p. 39, and Morton and Wess 1993, p. 348, where the concept of the experiment is credited to the Frenchman Jean Antoine Nollet.

33   Bugge wrote "Winkel Stratt", which would translate as 'shopping street', but Jan van Deijl (1715–1801) and his son Harmanus (1738–1809) lived in the Middelstraat or Vinkenstraat, parallel to the Haarlemmerdijk which leads from Amsterdam to Haarlem. Their optical workshop from the 1770s onward became the Dutch counterpart of the Dollond firm in London; see van Zuylen 1987 and Zuidervaart 2004, pp. 434-5.

A correct position of the ocular glasses is necessary to avoid the appearance of colours; for this purpose the colour rays have to be parallel and consequently to converge in one point (white light) at the back of the eye. He can also determine the correct position of the four given ocular glasses by geometrical construction. He showed me such new constructions for several of his telescopes, supports, etc. He charges no fixed price according to the length of the telescope, but charges according to its quality.

I have found the character of the Dutch to be such as it is generally described. At my inn, I saw them eat with their hats on in a party of 20 to 30 persons, and they began to smoke tobacco as soon as the sweet was served. The old and the middle aged maintain their simple manners and dresses; but as for the young and rich merchants, luxury is very much on the increase. *The new buildings which are erected in Amsterdam are in very good taste.* Mr Hasselgreen,[34] to whom I had been directed, was very polite to foreigners.

Those merchants in Amsterdam who are lovers of mathematics, and who have beautiful cabinets of instruments are: Mr Daniel Dornick, Mr Ebeling, and Mr van Noodt.[35] Lack of time, and also not knowing their addresses, made it impossible for me to visit them.

There are no public promenades in the town, except for the Plantage,[36] which is essentially just streets planted with trees. The places in between are used for buildings and gardens, which are regarded as the recreation grounds of the Dutch. Some small gardens have been fitted up as Vauxhalls. One evening I visited one of them, The Golden Ball;[37] there were no walking paths, no trees, poor music and singing, few lanterns in the garden, many small rooms without light, where the Amsterdam people sat down to drink or eat with their respective lady. It was all dreadful and seemed more like a brothel

34 In 1759, the Swedish astronomer Bengt Ferrner met a Mr Hasselgren in Amsterdam, who was identified as Jan or Carl Hasselgren, Swedish merchants in Amsterdam; Ferrner 1956, p. 83 and index. Presumably Bugge met someone from this firm.

35 Daniel Doornick's cabinet was sold by auction on 24 June 1778 in Amsterdam with a catalogue in Dutch and French, but no copy was found. A Dutch translation by him of a French publication on the electrical properties of tourmaline was published in 1773. Personal communication Huib Zuidervaart. Nothing was found on Mr. van Noodt. On Mr Ebeling, see below.

36 Bugge's "Plantagerne", which would translate as plantations, refers to the Plantage, laid out in the 17th century as an area with pleasure gardens.

37 A Vauxhall was a place of resort or amusement resembling Vauxhall Gardens in London, a popular pleasure resort from the 17th to the middle of the 19th century. Bugge already used this term in his report on Hamburg. We found no evidence in Amsterdam of such a venue named De Gouden/ Vergulde Bal or something similar.

August 1777
Amsterdam

than a decent and cheerful pleasure.

THE SLUICE
at Amstel Bridge.[38]

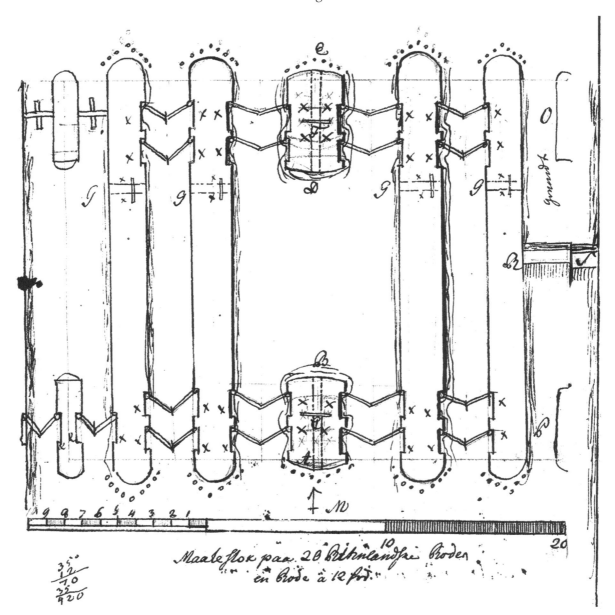

Measure of 20 Rhineland Roede, One Roede is 12 feet.

38   The Amstelsluizen is a complex drainage sluice, constructed in the river
     Amstel in the 1670s for the periodic refreshment of the water in the canals,
     and is still in existence today.

The whole arrangement of the great collection of 20 locks is easy to understand. Only it deserves notice that AB and DE are arched passages where the water outside at M or A and E is level with the water in the lock before the gates are opened. The small side locks have the same system of passages and gates in order to level the water in all of them.

At O and P walls have been erected, probably for the purpose of placing lock-gates there, as it has been done on the other side, but as the water is very shallow on this side, the project has been abandoned, and a brick dam has been erected from R to S.

Through Mr Hasselgreen I got to see the cabinet of Mr Ebeling, a solicitor.[39] This cabinet holds many beautiful objects; a complete apparatus for electricity; a small house, which can be destroyed by electricity

August 1777
Amsterdam

---

39 Ernestus Ebeling (1738–1796) owned a valuable cabinet, containing almost 500 instruments and models. For political reasons, he was forced to leave Amsterdam in 1787 and settled in Mainz in Germany. His collection was sold at auction in 1791, and among the buyers was Martinus van Marum. As a result, parts of his cabinet survive in Teyler's Museum, Haarlem. See Turner 1973 and 1973a and de Clercq 1994, pp. 33–34.

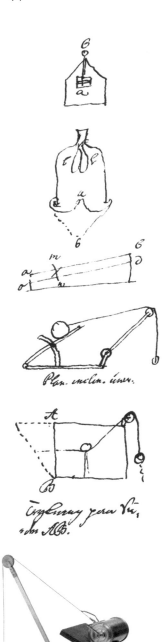

if the disk a is placed crosswise, or can be preserved, if it is placed below b. A beautiful air-pump with a complete apparatus by Nairne.[40] An experiment for showing the expansion and contraction of the lungs. A cross mn, which by its weight rolls down two strings ab and cd, but by electricity it is moved upwards. Two beautiful orreries, one for the Earth and another for the other planets, both made by Adams.[41] Two binocular tubes by Dollond.[42] A very complete microscope by Martin.[43] A 2 foot reflecting telescope with objective micrometers.

The simple machines. *[In the margin are two figures not discussed in the text, with captions]* [44]

An excellent model of the pile driver which was used in the construction of Westminster Bridge. It has been described by Desaguliers. It is operated with gears and can be used also with inclined piles.[45]

A model of a sawmill and several other mills. Several devices for demonstrating water pressure.

◀ *Binocular telescope by Dollond, London, length of parallel tubes 81.5 cm. University Museum Utrecht. Bugge saw a similar telescope in the private collection of Ernestus Ebeling.*

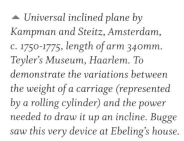

▲ *Universal inclined plane by Kampman and Steitz, Amsterdam, c. 1750-1775, length of arm 340mm. Teyler's Museum, Haarlem. To demonstrate the variations between the weight of a carriage (represented by a rolling cylinder) and the power needed to draw it up an incline. Bugge saw this very device at Ebeling's house.*

40 The London instrument maker Edward Nairne (1726–1806), whom Bugge was later to meet.

41 Famous family of instrument makers in London: George Adams senior (1709–1772) and his sons George Adams junior (1750–1795) and Dudley Adams (1762–1830), on whom see Millburn 2000.

42 The Ebeling sale in 1791 included an achromatic binocular telescope (Zuidervaart 1999, p. 522, note 60). Bugge's note shows that it was by Dollond and already in Ebeling's possession in 1777. His awkward description suggests that Bugge was unfamiliar with such a telescope, and although binocular telescopes had been made since the early 17th century (Bedini 1971), the achromatic binocular was indeed quite a novelty.

43 The author and instrument maker Benjamin Martin (1704–1782), whose shop Bugge would visit on 15 September.

A beautiful little machine, used by the Dutch peasants for scooping out the water. The machine is placed in the water.

On the 28th at 3 o'clock in the afternoon I went by canal barge through Haarlem[46] to Leiden, where I arrived at 10 o'clock in the evening. For me, my servant, and my trunk, I paid about 8 [...] in total;[47] the journey lasts 7 hours.

Mr Allamand, professor of physics,[48] was not at home. Mr van Wijnperse, professor of mathematics and philosophy,[49] is a handsome, ageing man. He showed me the Observatory of Leiden.[50] It is built on top of the Academy building, and it has been partitioned. The lower observatory is nothing but a long corridor with a window at one end towards the south. Here a transit instrument of about 2½ feet was placed. It was of rather old-fashioned construction, dating from the time of 's Gravesande.[51] Beside this instrument, placed on the pillar AB, was an equal altitude instrument like the one described in Smith's Optics.[52] It is rather good and made by Paauw.[53]

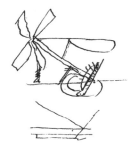

Leiden. I stayed at la Cour d'Hollande. Het Hof van Holland

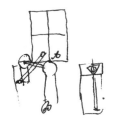

44 The captions are 'Universal Inclined Plane' and 'Pressure on side AB'. These two mechanics demonstration devices from Ebeling's cabinet survive in Teyler's Museum and the first is illustrated opposite; see Turner 1973, pp. 146-147 and 150–151.

45 This type of pile driver was invented by the watchmaker James Valoué and used in the building of Westminster Bridge from 1738 onward. A demonstration model was described and illustrated in Desaguliers 1747, vol. 2, pp. 416–18 and Plate 26, and it became one of the classic technological models in physics cabinets, see for example Morton and Wess 1993, pp. 107-109 and 149.

46 Bugge comments on Haarlem further on, fol. 31 recto.

47 The monetary unit here is not clear. The list of expenses (see appendix) has 'The journey from Amsterdam to Leiden 1 rdr. (rigsdaler)'

48 Jean Nicolas Allamand (1716–1788) had been a pupil of Willem Jacob 's Gravesande and succeeded Petrus van Musschenbroek as director of the university cabinet of physics in 1761.

49 Dionysius van de Wijnpersse (1724–1808) was professor of philosophy, mathematics and astronomy since 1768. He was far less involved with astronomy than his predecessor Johannes Lulofs.

50 Founded in the 1630s, it was housed inadequately on top of the University's main building along the Rapenburg canal, until a proper observatory was built in the 19th century for professor Frederik Kaiser. On the Leiden observatory in the 18th century, see van Herk 1983 and the relevant sections in Zuidervaart 1999 and 2007.

51 This transit instrument by Sisson, tube length 80 ½ cm, was bought in 1740 by 's Gravesande and is now in the Museum Boerhaave, Leiden.

52 The reference is to Smith 1738, §§ 844–851: 'Telescopick Instruments for finding time, by observing when the sun or any star has equal altitudes on each side of the meridian'.

53 In 1768 the professor of astronomy, Johan Lulofs, requested the construction under Paauw's supervision of an instrument for equal altitudes for the observation of the transit of Venus. This instrument is documented until the mid-19th century, but has not survived. Personal communication Huib Zuidervaart.

August 1777
Leiden

What I found very curious is that the transit instrument is removed from its bearings and placed in a case. Nor was there any spirit level to set the axis horizontally. In the lower observatory there is another room where several old and rather useless instruments are kept; a sector of 4 feet radius, old globes, planetariums, wooden quadrants; a 20 objective by Hartsoeker, 1686.[54] The best thing was an original Rhineland foot; however, the points of division were rather rough.

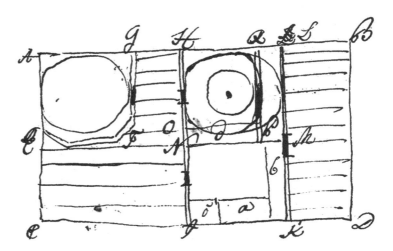

The upper observatory ABCD consists of the southern observation platform EFGHQP

View of the Academy building at Leiden. Engraving by A. Delfos, after a drawing by J.J. Bijlaerd, published in 1763. The platform with the astronomical observatory is half hidden behind the tower. ▶

where there is a roundish building EFGA[55] with a movable roof. It was intended to house a movable quadrant. *The roof is movable.* The construction of the support is rather unfavourable and unsteady. The quadrant itself was hidden in the house IKMN in a box a. Its radius was 1 ½ feet, and it was mounted with two telescopes, one of them movable. The edge was divided into whole degrees. And by wheelwork the degrees were subdivided into minutes and seconds, largely according to Hooke's methods.[56]

In the same building there was a 7-foot Newtonian reflecting telescope (b) by Hearne, London. It had two objective glasses and two plane mirrors, and Bradley's micrometer[57] in c, and in d several supports for the telescope. *The tube is of wood. The telescope is very good*.[58] Mr Wijnperse told me that at an elevation of more than 30 degrees it was difficult to make observations. The error is due to the support, which was, in a certain way, nothings but trestles.

OPHQ is a separate house with a movable (but leaky) roof, which houses an old-fashioned 4' quadrant.[59]

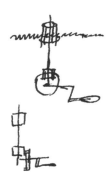

54 We cannot explain Bugge's '20'. The lens is in the Museum Boerhaave, Leiden. The name of its maker, the Dutch natural philosopher and optician Nicolaas Hartsoeker (1656–1727), and the date of production, 1686, are inscribed on it, but not (as on some Huygens lenses) the focal length. The lens has a diameter of 16.7 cm and a focal length of 1684 cm = some 54 Rhineland feet; van Helden and van Gent 1995, page IV. When the Swedish astronomer Bengt Ferrner was shown the same lens eighteen years earlier, he recorded it as "100 feet", which suggests that the correct focal distance was not known at the time. Kernkamp 1910, p. 471; a complete edition of Ferrner's travel journal is Ferrner 1956.

55 Bugge has EFGH.

56 This refers to the natural philosopher Robert Hooke (1635–1703).

57 The astronomer James Bradley (1692–1762), Royal Astronomer from 1742 to his death, took an active part in the improvement of instrumentation. Thus, in 1745 he added an improved micrometer screw to the 8-foot Graham quadrant at Greenwich to enable readings accurate to half a second of arc.

58 This wooden telescope, bought for Leiden University in 1736 by professor W.J. 's Gravesande, is illustrated below.

59 Made ca. 1610 for the Leiden professor Willebrord Snel van Royen (Snellius, 1580–1626) by the renowned Amsterdam cartographer Willem Janszoon Blaeu. The University bought the quadrant after Snellius' death as the first instrument for its new observatory. It is in the Museum Boerhaave, Leiden.

▲ *Reflecting telescope by George Hearne, London, 1736, length of tube 250cm. Museum Boerhaave, Leiden. Bugge saw it in the observatory and found it 'very good'.*

August 1777
Leiden

A 4 foot radius with a small azimuth Circle of about the following arrangement. *The quadrant is of wood, the edge of brass.*

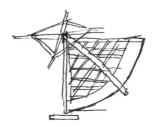

KDBLM is the northern observation platform. There are only two clocks and one of them goes only one day, and the other has no compound pendulum, and is moreover very bad in all respects.

The library is in another building. Its size is about ¹/₃ of the Copenhagen hall[60], and neither can it be large. There I found the second part of Hevelii Machina Coelestis[61] which contains observations. There are two catalogues of books, one subject catalogue and one author catalogue. Moreover, there are more old than new books.

In the library there were also two pairs of old globes, one pair by Adams. It had no hour circle, but on it was placed a brass wire parallel

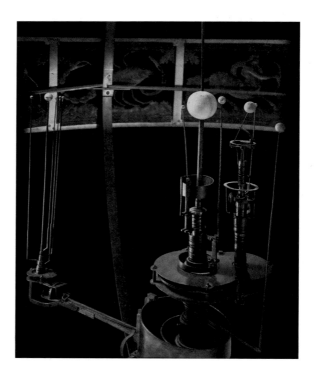

Planetarium by Steven Tracy, Rotterdam, c. 1670. Museum Boerhaave, Leiden. Inside the band depicting the zodiac, the planets move around the Sun in real time: Mercury, Venus, Earth with its moon, Mars, Jupiter and Saturn. At the time of Bugge's visit, these were still the only planets known. ▶

to the Equator which had a small movable index. The equator was divided into 24 hours. They are very neat all over. The diameter is somewhat more than 2 feet.[62] In a glass case there was a beautiful planetarium, and an armillary sphere which goes by clockwork. This is a very beautiful piece. The radius is about 3 feet.[63]

The next morning I visited Mr J.A. Fas. P.M. and Lector Matheseos, a very courteous man; he gave me his elements of integral and differential calculus, written in Dutch.[64] *On our way home he gave the address of Steenstra Lecteur en Astron: et Exam des Pilots a Amsterdam sur le Manege pres de porte de Leide.* [65]

I saw the botanical garden, which is very large and well-kept. The gardener Nicolas Meerburgh showed me a botanical work, published, engraved and illustrated by himself: Afbeeldingen van zeldsaame Gewassen. Leiden 1775.[66] He intends to continue this work. He knew Rotböl, Holm, and Brünich.[67]

August 1777
Leiden

60   Presumably Bugge refers to the Hall of Natural Specimens (Naturalie Sal), one of the five rooms in the Royal 'Kunstkammer' built in the 17th century.

61   The Danzig astronomer Johannes Hevelius (1611-1687) published his lavishly illustrated astronomical treatise *Machina Coelestis* in two parts in 1673–79.

62   This pair of globes by George Adams, diameter 18 inch, is listed in van de Krogt 1984 as ADA 8 and 15. Adams had given them to Leiden University in 1766; see Millburn 2000, p. 115.

63   This clock-driven planetarium, built ca. 1670 in Rotterdam by Steven Tracy, was bequeathed to the University in 1710 and placed in the library, where it could be seen in operation for more than a century. It is now in the Museum Boerhaave, Leiden. Dekker 1985.

64   Johannes Arent Fas (1742–1817) was appointed lecturer and extra-ordinary professor in higher mathematics and its application in physics and astronomy in 1763. The publication he gave to Bugge was Fas 1775.

65   Pybo Steenstra (?–1788) was lecturer in mathematics, first in Leiden from 1759 to 1763, then in Amsterdam. He published an introduction to astronomy, Steenstra 1771–72.

66   Nicolaas Meerburgh (1734–1814) had published the first part of his major work, *Afbeeldingen van zeldzame gewassen* (Portraits of Rare Plants) in 1775; it was followed by four additional volumes, and completed in 1780. The volumes contained fifty hand-coloured engravings of plants from the Leiden gardens.

67   The physician and botanist Christen Friis Rottbøll (1727–1797), the botanist Jørgen Tyge (Georgius Tycho) Holm (1726–1759) and the zoologist and mineralogist Morten Thrane Brünnich (1737–1827).

August 1777
Leiden

The natural cabinet is about half the size of that in Copenhagen, but everything is very clean and neatly arranged, at least to the eye; I am not able to judge the rest.

They also showed me a rubber boot; the skeleton of a traitor who would surrender Leiden to the Spaniards; he wore a wig and a Crispen collar; and a pair of the Emperor Charles V's boots.[68]

In Leiden there is a very able instrument maker, Paauw A.L.M.[69] He told me that he had concurred for the prize offered in Copenhagen concerning fire engines. He showed me a small engine with a pump cylinder, 1 foot high and 4 inches in diameter, which was able to spray with a remarkable speed and power over a house of two low floors. He asked me to send him Karsten's treatise.[70]

He makes very beautiful instruments, for example:
A collection of all sorts of pumps of excellent workmanship .... 260 florins.

68   A building in the botanical garden, the Ambulacrum, housed a collection of natural and other objects; see de Jong 1991. It is not clear which exhibit Bugge refers to. The inventories cited by de Jong list "a human skeleton", without details. Such a bizarre exhibit, a grim souvenir of the Spanish siege of Leiden in 1574, would not have been out of place among the rarities displayed in the Anatomy Theatre elsewhere in the city. But the inventories of that collection, reproduced in Witkam 1989, also do not list this particular skeleton. The 'Crispen collar' is equally mystifying; is there a relation with St Crispin, patron saint of shoe-makers?

69   The instrument maker Jan Paauw (c. 1723–1803) obtained the doctoral degree at the University of Franeker in 1762, after which he used the title Artium Liberalium Magister (A.L.M.) et Philosophiae Doctor. Rooseboom 1950, pp. 110–111.

70   On 30th May 1769, and again on 11th September 1770, the Royal Danish Academy of Sciences held a competion: "Give the best construction of a fire-engine in which cylinders, piston, tubes, valves etc. have the required strength, as well as the correct proportions according to the laws of hydraulics, and in which the handles obey the rules of mechanics, such that the whole machine can easily be transported through narrow alleys and extinguish the fire". Wenceslaus Johann Gustav Karsten (1732–87), professor of mathematics and physics at Bützow, later at Halle, was awarded a gold medal for his *Abhandlung über die vortheilhafteste Anordnung der Feuersprützen* (Greifswald, 1773). Personal communication Jørgen From Andersen, curator of Hauchs Physiske Cabinet, Sorø. See also Molbech 1843, p. 559.

A pyrometer .... 125 florins

This pyrometer consists of an oblong rectangular box ab, in which the rod is placed. The box is filled with water, which is brought to the boil by the lamps f.f.f. At g is an oblong cylinder, in which a thermometer is inserted; this indicates the degrees of temperature during the experiment, which is very important. But, on the other hand, it is not possible to test the extension of metals at temperatures higher than about 80° Reaumur.[71]

The Academy at Leiden has a separate collection of physics instruments.[72] The most remarkable were: three models of the fire machine, the first one is the very oldest pressure machine and it is of poor workmanship. The second has been made in England and is beautiful, but small and expensive (1000 florins). The third one is the most perfect and the largest. It has been made by Paauw for 1000 florins.[73]

There is a beautiful air pump

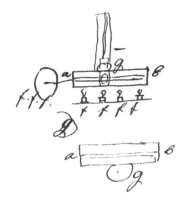

◀ *Pyrometer by Jan Paauw, Museum Boerhave, Leiden, 1759, height 63cm.*

▲ *Model of a steam engine after Newcomen by Edward Nairne, London 1772, height 79 cm. Teyler's Museum, Haarlem.*

71 A pyrometer, 'fire meter', was also called dilatometer, as it served to measure the expansion af metals when heated. A surviving example of the type of Paauw pyrometer that Bugge saw is illustrated above.

72 Physics demonstration lectures began to be held at Leiden University as early as 1675. Many of the instruments acquired for these lectures have been preserved and are now in the Museum Boerhaave, Leiden. See de Clercq 1997b, which describes and illustrates most of the objects mentioned by Bugge.

73 All three models survive. The steam engine after Savery, ca. 1730, and Paauw's steam engine after Newcomen, dated 1774 are in the Museum Boerhaave. The one 'made in England' was built by Edward Nairne in 1772, and is now in Teyler's Museum, Haarlem. de Clercq 1997b, pp. 168-170 and Turner 1973, pp. 184–185.

August 1777
Leiden

with a great number of bell-jars and apparatus for all kinds of experiments. Furthermore, in the theatre there is an old air-pump from 1675, made by Samuel Musschenbroek.[74]

Cylindrical and conical metal mirrors with requisite drawings. An excellently cut glass cone 2 to 3 inches high and wide.

Besides they have all the instruments described by 's Gravesande, as well as several new and beautiful ones, such as models of the Dutch water mills, a collection of Paauw's pumps; several other models of pumps and fire engines; a simple English pyrometer. A Belidor machine for moving marble blocks; a pile driver; cranes, etc., etc.

Once a week Mr Allamand gives a public lecture on physics in a theatre next to the instrument room.

Sunday, August 31.
The Hague

On August 31st in the morning I went from Leiden to The Hague and arrived there 3 hours later.

74   For this air-pump, made by the Leiden instrument-maker Samuel van Musschenbroek (1640-1681), see de Clercq 1997b, pp. 67–68.

Further it must be noted that outside Leiden there is a large number of bleacheries, and in the town itself there are several cloth mills. The wool was excellent, and I was told that the country produces its own wool.

The road from Leiden to The Hague is very pleasant; especially when you get near The Hague; on both sides of the canal there are very beautiful gardens.

The Hague is a very beautiful and tidy town. It has almost no canals apart from those around the town. Many private houses are very large; the Jewish part of the town is the most beautiful; all the houses there have uniform frontages. The Prince[76] has two palaces. Het oude Hof [The Old Court] is the lower, but now it is used only for the eldest succeeding prince. Het nieuwe Hof [The New Court], or the Prince's present residence, is an ill-arranged and unimpressive building, consisting of many

August 1777

The Hague
Het Engelse Parliament[75]

75 'Het Engelse Parlement', a rather upmarket inn in the street now named Korte Poten, was patronized mainly by English travellers. Personal communication Jan van Wandelen, The Hague City Archives.

76 William V (1751–1806), prince of Orange-Nassau, was the last stadholder in the Dutch Republic from 1766 until he fled to England when the French invaded the country in 1795. He was married to princess Wilhelmina of Prussia. Their eldest son was to become the first King of the Netherlands in 1815.

bits and pieces. The prince has one building, the princess a second, and their children a third one; the stable is in a fourth place.[77]

From The Hague to Scheveningen there is a very beautiful avenue with three lanes and mostly straight. Scheveningen is just a small fishing village situated on the dunes, where the inhabitants of The Hague promenade to amuse themselves with the view over the sea.

On the night of the 30th to the 31st August as well as the following day a violent onshore gale was blowing; and that day the water level at Scheveningen was so unusually high at 9 o'clock in the morning, that it was feared that the water would overflow the dunes and thus would have flooded a great part of the country, as it is below sea-level.

On Monday the 1st September, I went to Het Huis ten Bosch (Maison dans la bois).[78]

77   Bugge probably refers to respectively the Noordeinde Palace (now the work palace of the queen) and the Stadholder's Quarter or Binnenhof (now Houses of Parliament).

78   In the 1640s Princess Amalia van Solms erected a single-room edifice, the Oranjezaal, outside The Hague as a summer residence and had it decorated with a set of 31 paintings as a mausoleum for her late husband, Stadholder Frederik Hendrik. Between 1730 and 1754, it was extended and made fit for permanent residence and became known as the House in the Wood. At Bugge's time, it was used by Stadholder William V and his family, who added a Japanese and a Chinese room. Since the 1980s it is the main residence of the queen. Loonstra 1985.

The road goes through the menagerie, where there are roe deer and a few wild boars. The road is an avenue and very comfortable.

The palace is very small, but nice. The most beautiful rooms are 1.) The Chinese room. 2.) A room with: a chandelier, two-armed candelabras, a clock case, a great number of large and small vases, everything made of very beautiful Berlin porcelain. This is a present from the King of Prussia to the Stadholder's wife. 3.) Family rooms with life-size portraits of the elder Princes of Orange and their consorts. 4.) The Oranjezaal is very beautifully painted, and as they said there were pieces by 9 different masters. The most beautiful are: the triumph of the Prince of Orange, painted by J. Jordans. Venus handing to Mars his weapons, by Tylde [Van Thulden], and the two smirking[79] Cyclopes, by Rubens. It is a great masterpiece.[80]

September 1777
The Hague

GEZICHT van het BUYTEN HOF
Opgedragen aan den Wel Edele Gestrenge Heer
Mr CAREL DE LA BASSECOURT,
Regerent Scheepen van 'S GRAVENHAGE. 1758.

VUE & PERSPECTIVE du BUYTEN HOF
dedie à Mr CHARLES DE LA BASSECOURT,
Echevin Regnant DE LA HAYE. 1758.

◄ *The Buitenhof in The Hague, etching published by H. Scheurleer in 1758. On the right the Binnenhof, the Stadholder's Quarter, with the observatory on the tower, which surprisingly Bugge does not mention. On the left the Stadholder's museum, where Bugge saw the natural history collection.*

79 Bugge wrote 'smidskende' which means 'smirking', but that makes no sense in this context. We presume a connection with the word 'smith', and that he meant 'forging', which is what the Cyclopes in the painting are doing.

80 Bugge singles out three paintings: the large Frederik Hendrik's Triumph by Jacob Jordaens (1652) and two smaller paintings after scenes from Virgil's *Aeneis*, viz. the smithy of Vulcan (1649) and Venus selecting Aeneas's armour in the smithy of Vulcan (1650), both by Theodoor van Thulden, a pupil of Rubens. The smithy of Vulcan (a scene also painted by Rubens, but that was not the one Bugge saw) shows three Cyclopes hard at work. Peter-Raup 1980.

September 1777
The Hague

The prince's cabinet of curiosities, his collection of specimens of animals, flowers, etc., as well as his library and instruments *and paintings* are in a beautiful building behind Het Oude Hof.[81] As a matter of fact, the cabinet of curiosities has only two rooms; but its collection of East and West Indian dresses, weapons, models of their vessels and houses, and furniture is extremely magnificent.

The natural history cabinet consists of 5 rooms; *the first* houses butterflies and insects; *the second*: minerals and a variety of stones, corals, etc.; in the centre stands a very fine stuffed sea lion;[82] *in the third* room there are snails, fishes and snakes in spirits, and some animals; *the fourth* houses animals and birds, and an orangutan in spirits which used to live in the Prince's menagerie; the fifth houses

81 The stadholder's 'museum' was housed in a building on the Buitenhof, some fifty yards from the Stadholder's Quarter or Binnenhof, not 'behind the Oude Hof' as Bugge reports. The collections included instruments, which Bugge evidently did not see. Surprisingly, he also does not mention the stadholder's observatory, which had been erected on a tower of the Binnenhof in the 1750s. See de Clercq 1988.

82 From other descriptions of the stadholder's 'museu' it is known that the centrepiece in the second room was a stuffed hippopotamus. In 1798, Bugge visited the National Museum of Natural History in Paris, which contained among others the stadholder's naturalia that had been carted off by the French three years earlier: "Here I had a second view of some singular objects, which I had seen at the Hague one and twenty years before, in the Stadtholder's collection, such as the sea-horse, zebra, elephant, orangoutan, and a variety of monkeys". From the 1801 English edition of Bugge's Paris journal, as quoted in Crosland 1969, p. 69. An antiquated meaning of seahorse is hippopotamus; the Danish edition has the normal term 'flodhest'.

birds; in the centre there is a stuffed orangutan.[83] What especially pleased me was that all animals and birds were properly and naturally stuffed and were placed in very well-chosen postures.

In The Hague there is a French theatre. The house is very small; but the actors are rather good.

On my passage through Haarlem I saw nothing extraordinary except the church of Haarlem. It is a very large church, and its organ is most extraordinary. It is said to be one of the most beautiful in the whole of Europe.[84]

From The Hague you go back[85] by canal barge in order to get to Delft. There I saw the church, where the Princes of Orange are buried *it is large but built in the Gothic style*. The burial place is a small square area enclosed with iron fencing. There was only one monument. At the side

September 1777
The Hague

The 2nd September

83   In 1776, the director of the stadholder's cabinet of natural history, Arnout Vosmaer, acquired the first living orang-utan ever to arrive in Europe. It died in the stadholder's menagerie in January 1777 and Vosmaer had it stuffed, which led to a quarrel with his colleague Frans Hemsterhuis, who had promised the animal for dissection to the naturalist Petrus Camper, if it were to die. The Stadholder personally intervened in the quarrel. Pieters 2002; the episode was first reported, with full transcript of the correspondence, in Mazel 1909.

84   Bugge only now writes what he had seen four days earlier when he passed through Haarlem on his way from Amsterdam to Leiden. He probably did not get off the barge and only had a distant view of the church, with its famous organ built by Christian Müller in the 1730s.

85   The Hague was linked to the canal leading from Leiden to Rotterdam by a spur canal. For maps and other information on the canal network in the Republic, see de Vries 1981.

September 1777

other men of means are buried. Somewhere else in the church was a beautiful marble monument for a Dutch admiral.[86]

In Delft there are many factories for the so-called delftware or blue and white stoneware. I saw two of them. Part of the clay comes from the other United Provinces, but part of it also comes from Germany.

The 3rd September

From Delft I went by canal barge to Rotterdam. Particularly the bourse in Rotterdam is a very beautiful and nice building, dressed with sandstone. It is larger than the bourse in Amsterdam. The town is tidy and neat. It has many canals, two of which are so deep that very large ships are able to enter the town. Generally there is a shipping service from Rotterdam to Hellevoetsluis, but because of the wind and the high tide I had to go by coach; I had to change twice. *The road went through Brielle, a small and beautiful fortified town.* At 3 o'clock in the afternoon I arrived

86 Bugge mentions only one church, but he must have visited both the Old and the New Church, as was customary (Scholten 2001, pp. 211–231: "Epilogue: tomb tourists and funerary antiquaries"). The famous monument for William of Orange – nicknamed 'the Silent' – by Hendrick de Keyser (1614–22) is in the New Church, whereas Bugge's reference to a monument for a Dutch admiral must be to the tomb of Maarten Tromp by Rombout Verhulst (1658) in the Old Church.

at Hellevoetsluis; but due to storm and contrary winds the packet boat could not leave until the 4th at 9 o'clock in the morning. We had no storm during the passage, but only little wind and contrary wind; so that we did not arrive at Harwich until the 6th at 5 o'clock in the afternoon.

Hellevoetsluis is a small fortified town. In its centre is an excellent harbour with 8 to 10 Dutch war ships with the requisite masts and storage rooms. Because of the tides, by means of a very big sluice ab, the ships can easily be careened and the harbour cleaned out, as the sluice is closed at high tide, but opened at low tide.

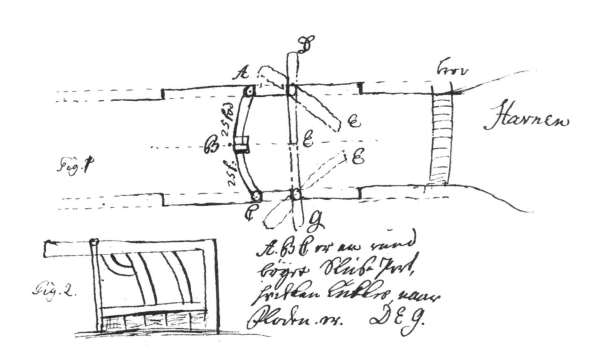

ABC is a round, curved lock gate, which is closed at high tide. DEG

September 1777

are two other gates, which are closed at low tide in order to keep the warships in sufficiently deep water. Fig. 2 shows a drawing of the lock gate DE.

As there are no stage coaches on Sundays, 6 of the passengers teamed up to hire a carriage to London. My share was 1 pound. At 7 o'clock in the evening we left Harwich, and arrived in London on Sunday the 7th at 11 o'clock in the morning

London, the 7th
Great Suffolck Street No. 31.
near the Hay Market by
Mr Bower, Taylor. [87]

Chelsea Hospital or College was one of the first things that I saw.[88] It is a very large and beautiful building for the disabled, the plan is more or less as follows

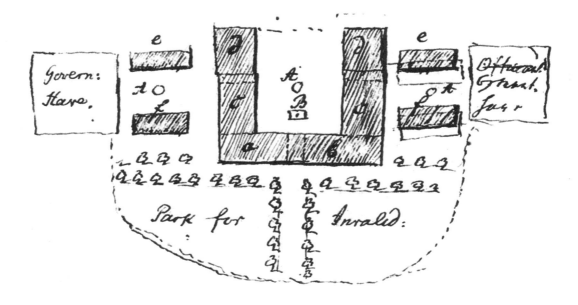

---

87 Evidently this is where Bugge took accommodation.

88 The Royal Hospital in Chelsea was founded by Charles II in the 1680s as a home for army veterans and is still in use as such. Walker 1987, pp. 128–131.

a. the chapel b. the dining hall; they have meat 3 days a week: c.c. the room of the disabled. d.d. the two Governors' and the officers' rooms. e.e. dispensary, barber-surgeon and ward. f.f. private wards, kitchen, guardroom etc. A.A.A. are portrait busts placed within beautiful iron fencing. B. is a statue of King Charles, founder of the hospital.

Kensington Garden and Palace is a mediocre building. The garden is large, according to English custom; it consists of avenues and terraces.

Ranelagh[89] is a place where, from Easter to July, people enjoy themselves with dancing. In the centre is a fireplace that reaches all the way to the ceiling. b. is the orchestra stand. In a circle c.d.e.f. around the fireplace are placed tables and benches for eating and drinking. On the outside it has equally been divided up into many boxes on the first as well as on the second floor.

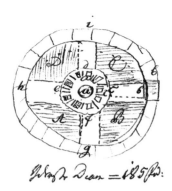

Outer diameter = 185 feet

89   Ranelagh Gardens, east of Chelsea Hospital, were laid out as pleasure
     gardens in 1742. The centrepiece was a rotunda designed by William Jones,
     one of the largest buildings of London at the time, larger than the Pantheon
     in Rome. It was demolished in 1805. Walker 1987, pp. 136–140.

September 1777
London

cb and di and ef and fg are places where they dance. A,B,C,D are covered with straw mats. The profile of the hall is as shown.

The hall is illuminated by 38 chandeliers hanging in two concentric[90] circles. Furthermore, all cornices and columns are hung with lamps, about 4000 in total. *The paintings, hanging in the cornices, are burlesque and in very bad taste. The fireplace in the centre obstructs the view. Nearby there is a nice garden which is also illuminated.*

The largest fire machine in London is just outside Chelsea.[91] It corresponds to Belidor's drawing.[92] It makes 16 strokes per minute and

---

90   Bugge wrote parallel circles.

91   The Chelsea Water Works Company pumped up Thames water to supply it to the citizens. For this purpose it installed two steam engines in the 1740s, which were objects of great curiosity at that time. Walker 1987, pp. 141–143.

92   Bernard Forest de Bélidor (1697/8–1761), engineer and writer on mechanics and engineering. Bugge probably refers to the engravings of a Newcomen steam engine in Bélidor 1737–39.

in that time it pumps up 5 hogsheads of water.[93] There are two machines placed side by side, so that one can work while the other is under repair or is short of water. It works day and night on all weekdays.

The Foundling Hospital[94] is a very nice building, and the indoor arrangement is very tidy and neat. The children looked very healthy. At that time there were about 120 of both sexes. It was said that about the same number were in the country for their health.

The girls' winter dining hall is decorated with a portrait of the founder (an East Indian captain 1739), but besides there are life-size portraits of many other benefactors, e.g. Doct. Mead,[95] King George II etc. Plaques stating the value of the donation and the names of the donors are hung up. The boys' dining hall is also very beautiful.

Several other rooms for the Governors are very nicely decorated

93 Bugge writes that the machine 'bruger' ('uses') the stated quantity of water, by which surely he means 'pumps up'. According to Barlow 1740, "8 cubic foot of water make a Hogshead and 4 Hogshead a Ton". So according to Bugge, the machine pumped up some 1250 litres per minute.

94 The Foundling Hospital for abandoned children was founded in 1739 by the philanthropist Captain Thomas Coram. The painter William Hogarth (1697–1764) promoted the scheme and presented a portrait of Coram. Other artists also gave paintings to decorate the Governors' Court Room, which brought income for the hospital through paying visitors, so that this was effectively the first public gallery in London. George Frideric Handel also supported the hospital's charitable work by giving benefit performances of his work in the chapel. The organisation continues to this day as the Thomas Coram Foundation for Children, with the art collection and parts of the interior now on display in the Foundling Museum, Brunswick Square. Nicolson 1972.

95 Richard Mead (1673–1754), physician and collector of books and objects of art.

September 1777
London

with historical paintings. One room with historical scenes from the New Testament, painted by an English clergyman.[96] Another room with histories from the Old Testament.

In a western room is a painting by Hogarth; it is called Hogarth's masterpiece. It represents the march of the English Guard Regiment against the Scottish rebels.[97] Two women, one with a child and another who is pregnant, quarrel over a soldier. One soldier is drunk and falls on the ground; one of his companions wants to give him some water, but he reaches for a glass of brandy which a sales apprentice offers him. Another soldier overturns a girl's milk pail and lets it run into his hat. A little chimney sweep holds his black skullcap underneath. Another soldier grabs a girl under her skirts, as she had climbed up in order to watch two men boxing.

The chapel is very nice; in the middle of the floor is a fireplace of cast iron with bronze decorations shaped as

▲ *The March of the Guards to Finchley by William Hogarth, c. 1749, oil on canvas, 101 × 133cm. Foundling Museum, London. It shows a fictional mustering of troops to defend the capital against the Jacobite rebellion of 1745, which aimed to return the Stuart Dynasty to the throne.*

a cylindrical pedestal with a vase of very tasteful workmanship. The smoke escapes underneath the floor.

The Pantheon is one of the buildings used for balls and masquerades. It was built 6 or 7 years ago. It is a private estate.[98] The entrance fee is ½ crown or 2 ½ shilling. The large hall goes right through the building and is illuminated by the cupola abcd or rather the lantern e. The decorations are placed neatly and very tastefully, and consist of patterned marble pillars, columns and statues. Furthermore, at AB and CD is a corridor between the columns, provided with benches and canopies.

On the second floor from A to B and from C to D is a beautiful gallery. On the same floor from A to C and from B to D are dining rooms, small tea and coffee rooms. The basement of the house has been fitted with two ranges of tables and benches for the same use. The entrance fee for the masquerade is 2 guineas, but then you are allowed to eat and drink as much as you like from 9 o'clock in the evening till

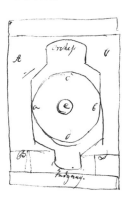

Pantheon

96   James Wills, who painted *Little Children brought to Christ* in 1746, took up a curacy in Leicestershire in the 1750s. Nicolson 1972, cat. nr. 82 and p. 26.

97   For a discussion of this painting, see Uglow 1997, pp. 419–425.

98   The Pantheon in Oxford Street, designed as a "winter Ranelagh" with a similarly impressive rotunda building as the one in Chelsea that Bugge visited earlier, was opened in 1772 to general acclaim. Rebuilt after a fire in the 1790s, it was demolished in the 1930s to make place for the department store Marks & Spencer.

99   Following advice of a Royal Society Committee, the Board of Ordinance, to protect its gunpowder magazines from lightning, had set up pointed metal rods above the roofs, according to Benjamin Franklin's theories. But on 15 May 1777, their premises at Purfleet in Essex were struck by lightning. One committee member, Benjamin Wilson (1721–1788), FRS, a keen explorer of static electricity, had recommended rounded rods below the level of the roof instead, and King George III now provided him with the means to test his theories. Wilson set up an 'artificial cloud' or prime conductor, 155 feet long and 16 inches in diameter, in the Pantheon. A scale 1:36 model of the Purfleet arsenal armed with pointed or blunt conductors raised to diverse heights was drawn on rails beneath the prime conductor to simulate the relative motion of clouds and building. See Wilson 1778, and Heilbron 1979, pp. 381–383.

100 Bugge's 'Runde Træ Spaaner' here translated as 'round slips of wood' after Wilson's own description (Wilson 1778, p. 251): "To construct the substitutes for a cloud, I first joined together, in 15 lengths, the broad rims of one hundred and twenty drums (merely to have them portable) by means of wood cut into long slips, which were fixed on the inside thereof [...] covered with tinfoil".

101 Capacitators, better known as Leiden jars after the town where Peter van Musschenbroek accidentally discovered their properties.

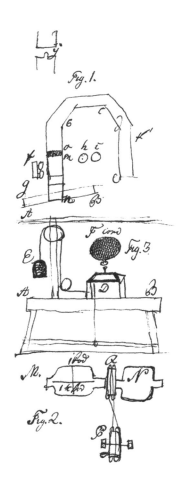

11 o'clock the next morning.

Mr Wilson had set up a complete electrical apparatus in the Pantheon [99] abcd is a continuous conductor made of round slips of wood[100] covered with tinfoil. mn is another shorter conductor of the same diameter, 1 foot, and about 7 feet long. This conductor is not really connected with the former, but when it is charged, a spark goes from the spigot y. f and g is the electrical machine consisting of two cylinders M and N, about 1 foot in diameter and 1½ feet long; they were set in motion by the wheel P and the string PQ.

AB, which is shown in profile in Fig. 3, is just a sledge over which the house D can be moved. When it reaches the point B, it is carried back towards A by means of the weight E. The passing of the house under the conductor F represents a cloud passing over a house: h and i are two big Musschenbroek jars[101] which are used for other experiments.

Electrical experiments at the Pantheon, London. Engraving from B. Wilson, 'New experiments and observations on the nature and use of conductors', Philosophical Transactions of the Royal Society of London 68 *(1778). In London, Bugge witnessed these truly spectacular experiments to determine the best shape of lightning conductors.* ▶

From what people said I gathered that Mr Wilson intended to prove two things.

1° that the pointed conductor on top of the house a is not as good as the ball-shaped conductor b.

2° he claimed to be able to ignite powder without fire, or rather, that the electric fluid alone sets the gunpowder on fire.

Among those present was Lord Mahon[102] who raised very witty objections to Mr Wilson's claims.

Bridewell are prisons for lecherous women and other criminals, where they have to wash, spin or do other work according to circumstances.[103]

The 13th September

Mr Forster[104] took me, Doctor Tetens,[105] Mr Gulden, and the Court Councillor Born,[106] to a painter, Rigaud[107] and Oberien [= Cipriani?][108], whose paintings were most excellent. A sketch for a piece from English History, representing an entry after a victory over the French. Jupiter making one of Diane's nymphs pregnant disguised as a Goddess;

102 The politician and inventor Lord Mahon, later 3rd Earl Stanhope (1753–1816), FRS, devised among others a new printing press which bore his name, and a type of lens later used in the manufacture of novelty souvenirs named Stanhopes. For his electrical theories and his ideas on the proper shape for lightning conductors, see Heilbron 1979, pp. 462–464.

103 There were three prisons of that name in London: on the banks of the Fleet River, at Westminster and at Clerkenwell. It is unclear which, if any, Bugge visited.

104 Probably the naturalist, ethnologist, travel writer, journalist and later revolutionary, (Johann) Georg (Adam) Forster (1754–1794), FRS.

105 Possibly Johann Nicolai Tetens (1738–1807), since 1776 professor of philosophy, later also of mathematics, in Kiel. He had studied in Copenhagen and may have been a fellow-student of Bugge.

106 Possibly the Viennese mineralogist and metallurgist Ignaz von Born (1742–1791), FRS. Bugge calls him "hofraad"; in Austria, Hofrat was the standard title for a high civil servant. In 1776, Born had been appointed by the empress Maria to arrange the imperial museum at Vienna, where he was nominated to the council of mines and the mint.

107 The Italian history and decorative painter John Francis Rigaud (1742–1810) settled in London in 1771. Rigaud had entered the paintings mentioned by Bugge at the Royal Academy Exhibitions of 1772 (Jupiter, under the form of Diana, visiting the nymph Calisto, after Ovid), 1773 (Cupid sharpening his arrows) and 1774 (The Entry of the Black Prince into London with his Royal Prisoner). See Rigaud 1984.

108 Our initial thought that Oberien stands for O'Brien proved a dead-end. Instead, we suggest that 'Oberien' (Ciberien?) is a garbled version of Cipriani. Rigaud at times worked together with the Florence-born decorative painter and draughtsman Giovanni Battista Cipriani (1727–1785), who in London was "one of the great backroom figures of the neo-classic style in England". In September 1777, they worked for several weeks on the proscenium and ceiling of Covent Garden Theatre, see Rigaud 1984, pp. 20 and 62.

September 1777
London

Cupid sharpening his arrow, besides several well executed portraits.

From there we went to Wedgwood and Bentley.[109] At ground level is the shop of the ordinary Queen's Wares or yellow flints. On the second floor are objects of the same kind but decorated with blue, green, reddish, and especially a cinnabar colour of recent invention. In the cellar of the same shop Staffordshire ware. Everything was very beautiful, but extremely expensive. A cup and saucer £3. In another room were copies of all Greek and Roman earthenware and stone works, such as urns, vases, bas-reliefs. Herculanum drawings on flat stones, busts, etc. I bought a bust of *Reinhold* Forster[110] and paid half a guinea for it.

The 15th [September]

I went to see the instrument maker A. Smith, Strand near Charing Cross, at the Golden Quadrant.[111] He is a very polite man. At his shop I bought a ruler with English, French, Dutch, and Antwerp foot measures for 3 shilling. And I ordered a triplet prism in order to prove

109 In 1774, the Staffordshire potters Josiah Wedgwood and Thomas Bentley opened showrooms in Portland House, 12 Greek Street, near Soho Square. In the 1770s they published two catalogues: one of Ornamental Wares, the other of Queen's Ware. Reilly 1995, pp. 96 and 387–389.

110 The Prussian naturalist and philosopher, Johann Reinhold Forster (1729–1798), accompanied Captain Cook on his South Sea Voyage in 1772. He was the father of the aforementioned Johann Georg Adam Forster. In 1776, he was modelled by Joachim Smith for Wedgwood, who also brought out a portrait medallion of him, illustrated in Reilly 1973, n. 33.

111 Addison Smith, optical and philosophical and mathematical instrument maker, documented working period 1769–1789, see Clifton 1995, p. 254. A trade card, reproduced in Crawforth 1985 as fig. 48, gives as his address: "Opposite Northumberland Street, Strand, London". The shop name given by Bugge, 'At the Golden Quadrant', was not previously known.

experimentally the nature of Dollond's telescopes. He promised to bring me in contact with Mr Dunn. He told me that, long before Dollond, Mr Hall, a gentleman and amateur, had begun to combine two pieces of glass in order to cancel out the colours and that he had taken the idea from Newton's Opticks 1st Book, VIIth Proposition.[113] On the whole Mr Smith's work does not seem to be very remarkable.

Benjamin Martin, Fleet Street No. 171. There I bought some books.

At Dollond's, St. Paul's Church Yard[114] I bought a compound microscope for 8 guineas. In addition I got several of his books.

Finally at Nairne and Blunt's, near the Royal Exchange.[115] This splendid man has many excellent objects. A new type of marine barometer which does not change its height in spite of the movements. a is a pivot, movable through the hole in AB, and thus the barometer can move to one side. The circle aghi is connected to this pivot.

Catal. Inst.[112]

Cat. Inst.

Cat. Inst.
Cat. Inst.

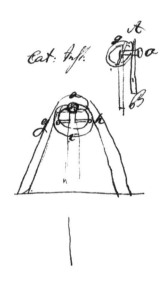

112 Catal. Inst. in the margin, here and elsewhere, suggests that the instrument maker gave Bugge a copy of his trade catalogue. No trade catalogue by Smith is known to have been preserved.

113 On the dispute over the true inventor of the achromatic lens, Dollond or Chester Moor Hall, see Sorrenson 2001.

114 Peter (1731-1820) and John junior (1746–1804), the sons of John Dollond senior, had gone into partnership in 1766. Peter ran the shop in St. Paul's Church Yard and John that in the Haymarket, which Bugge would visit nine days later.

115 Edward Nairne (1726–1806) went into partnership with his erstwhile apprentice Thomas Blunt in 1774. For an overview of Nairne's career and products, see Warner 1998.

London
September 1777

Cat. Inst.

Cat. Inst.

The barometer is attached to gh like a hanging compass, and can thus move to the other side.[116]

A very beautiful inclination compass of 6 inches radius, on a foot which could be revolved and placed horizontally by means of two spirit levels. Many sorts of theodolites; some of them could not be verified; others were very well arranged with all appropriate movements.

A new kind of support for large telescopes; two springs are pressed down, and then the movement is free. If they are pressed upwards the movement stops at the fine screws.

Furthermore, I visited several other instrument makers, namely:

Samuel Witford No. 27 Ludgate Street near St. Paul's Church-Yard. This man seems to be of the ordinary kind.

Benjamin Martin & Son; Fleet Street No. 171; he has a well-stocked shop. I bought some of his books.

116 In the search for usable maritime barometers, which would work reliably on a moving ship, Edward Nairne devised a constricted-tube barometer, fixed on gimbals with a weight at the bottom to keep it upright even when the ship was rolling. It was first used during Captain Cook's second voyage of 1772–75, and Nairne's barometers were soon taken on other British exploring voyages. See McConnell 2005 and Middleton 1964, p. 163.

I went to see the very beautiful Hospital for Disabled Seamen at Greenwich.[117] Everything is clad with Portland stone. The chapel is extraordinarily fine, and the placing of the altar in a very large niche has a very good effect. The two big halls[118] painted in fresco are very beautiful. A description of them is sold at 6d. The disabled eat in a very big hall proper and neat. Besides they generally sleep six in each room; and I think that clean and fresh air in the hospital is not sufficiently taken care of. I was told that foreign sailors were received in the hospital in preference to Englishmen. A good policy encouraging foreigners to enrol.

Besides the disabled there are furthermore 100 poor boys, sons of sailors living in England. They are educated in mathematics, navigation, and maritime affairs.[119]

At Woolwich I saw[120] the whole arrangement of the laboratory, the foundry and the boring machine[121]

The 16th and 17th [September]

117 Established in 1694 for the relief and support of seamen and their dependants, and built to the designs of Christopher Wren. The ceiling and wall paintings in the Great Hall were executed by James Thornhill between 1707 and 1726. The chapel was to be destroyed by fire two years after Bugge saw it, but was rebuilt. Between 1873 and 1998, the complex served as Royal Naval College. See Bold 2000.

118 The Great Hall has two sections, Lower and Upper Hall.

119 Instruction had begun in 1715 when Thomas Weston, who had worked with the Astronomer Royal, John Flamsteed, established his Academy. See Bold 2000, chapter 8.

120 In his Paris journal, Bugge describes the Arsenal, and states 'Neither in France nor in England is any mystery made of the constructing of the drilling machine. When in 1777 I met the English artillery-captain, Smith, I was given permission to visit the foundry and drilling-machine as often as I wished'; quoted from Crosland 1969, p. 147.

121 Situated east of Greenwich, this was since the later 17th century the location of the military complex Woolwich Warren, named Royal Arsenal in 1805. The buildings included the Royal Laboratory for the production of ammunition, and the Royal Brass Foundry for the production of cannons and mortars. In 1770, Jan Verbruggen (1712–1781) was appointed Master Founder. With his son Pieter (1735–1786), he introduced horizontal boring for guns cast solid. Previously guns had been cast around a core and reamed out vertically. In his boring machine, the borer-head remained stationary, the gun being made to revolve by four horses. See Hogg 1963, p. 1416 and p. 1028 (photo of a model of the boring machine in the Royal Artillery Museum). For a series of fifty coloured drawings showing the operation of this gun foundry in the 1770s see Jackson and de Beer 1974, and especially de Beer 1991.

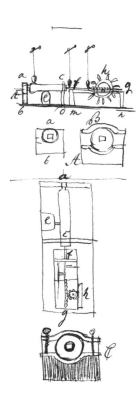

through the assistance of Captain Molman, who lives there.[122]

The arrangement of the boring machine. The cannon is revolved by means of a wheelwork driven by horses in A. It is fastened to ab as shown in B. The muzzle can be revolved at cd as shown in C. These blocks are of metal. The drill itself is lying on tables mfgn. At one end it has teeth, so that it can be moved forward or backward by means of the wheel h.

The handle was operated in the following way: a man pulled it a bit forward, stopped it for a while, and then drew it back again. The chips are about 1 inch long and thick. E is a chair which can be moved forward and backward, in order to place the turning tools so that the cannon can be turned and drilled at the same time.

▲ *Gun boring machine in Woolwich Arsenal. Coloured drawing, attributed to Pieter Verbruggen. In Woolwich, to the east of London, Bugge visited the gun foundry and inspected the boring machine. As the gun barrel was rotated by the power of horses treading outside the building, the stationary drill bored it out to the required diameter.*

The cannon founder Verbruggen is a Swiss,[123] but a very polite and courteous man and a great master of his art. His cannons were bored like a mirror, and no air bubbles were visible neither inside nor outside. He told me that he used no bellows in his melting furnace, but only a number of draught pipes placed at various directions according to the wind direction. In six hours he could then melt just as much metal and just as well as others could do in 24 hours.

Master S. Dunn,[124] Teacher of Mathematics, came to see me; he complained about Maskelyne's[125] behaviour against him. He told me that he had found 1º) a method of finding the longitude by means of the declination of the magnetic needle. 2º) His theory allowed him to find, from observations of two declinations, the deviation in all intermediate places.

London
September 1777

122 Captain John Mollman retired from the Royal Artillery in 1765 and was appointed Assistant to the Chief Firemaster in the Royal Laboratory. Hogg 1963, p. 327.

123 This is not correct: the Verbruggens (Bugge wrote 'Vanpruggen') came from Enkhuizen in Holland.

124 Samuel Dunn (1723–1794), writing master, teacher of mathematics and navigation, astronomer. He observed the 1769 Venus transit at Greenwich with Maskelyne, and introduced a new method of finding longitude; see Maskelyne 1778 and Wallis 1986, p. 340.

125 Nevil Maskelyne (1732–1811), Astronomer Royal since 1765, had residence at the Greenwich Observatory, which he would later show to Bugge. As a member of the Board of Longitude, he was heavily involved in all proposals to find longitude, in which context he had clashes with clockmakers. In 1774 he was awarded the Royal Society's highest award, the Copley medal, for his work on the density of the earth.

September 1777
London

The 20ᵗʰ

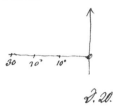

He has verified this by consulting the journals of the East India Company; 3°) that a young English naval officer, namely Townshand, during several voyages and repeated trials has discovered that, knowing the distance from the moon to the fixed stars, it is possible to determine *the longitude within* 10 Leagues, or English or French nautical miles, of which 20 = 1° circ. max. By this method the error does not increase in proportion to the distance from the meridian, as is the case with sea clocks.

I visited Mr Nairne who showed me several experiments with the electrical machine. 1). Dancing paper pictures. 2). Ringing bells. 3). That the electric atmosphere reached beyond his room, about 12 feet from the electrical machine. 4). When trying to remove a guinea from a portrait covered underneath with tin foil you received an electric shock. 5). The procedure of charging with positive and negative electricity. A is a ball, BC is the stick

Trade card of Edward Nairne, c. 1760. Science Museum, London. These forerunners of the present-day visiting card gave prospective customers an idea of the range of goods an instrument maker could supply. ▶

which has now been placed sideways. CD is a conductor which does not touch the ball, but the rear part of the stick. EF is another conductor. G and H are two Leiden jars.

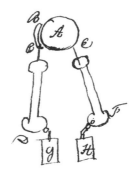

When the ball is revolved, G is negatively charged and H positively. a.) G − and H + together produce a spark. b.) If they are both charged at DC or at EF, then they both become either + or -, and they produce no spark, but the spark is produced in the usual way. c.) If they are both charged G − and H + at a certain number of revolutions, say 20, and they are then exchanged so that H is put in G's place and G in H's place, then all the electric substance is removed. d.) The powder magazine is ignited by a 9-jar battery.

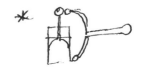

This is a small house which can fall apart with hinges.[127] bcde is a brass plate connected to a by a steel wire. fg is another plate connected to the electric handle by the steel wire gh. A small rocket is placed at e. It is ignited by the electric

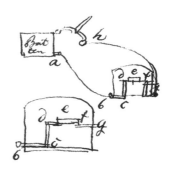

126   We have been unable to identify this Townshand. His name and alleged accomplishment are not mentioned in the proceedings of 'The Longitude Symposium', held at Harvard in 1993; Andrewes 1996.

127   The thunder-house was a standard -piece of electrical demonstration; see for example de Clercq 1997b, p. 150. Usually, a container with gunpowder placed between a spark gap of brass wire was used. The captions to the diagram are 'brass / powder / brass'.

September 1777
London

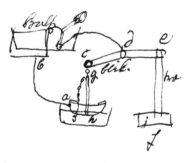

The 21st and 22nd.

spark and causes the house to fall apart at the hinges.

e.) The preservation of ships. The ship is connected to the battery by the steel wire ab. Fig cdef is a tin conductor placed just above the mast. This was of glass, and to its upper end a piece of round steel wire had been fixed by means of putty. As long as the chain hangs from g down into the water, the mast remains intact during explosion. As soon as the chain is removed, the mast is broken to pieces.[128] f.) The aurora borealis in a straight cylinder as well as in an arch, both evacuated. g.) to brand gold into glass. h.) objects with Canton's Phosphorus.[129]

At his iron factories Wilkinson[130] no longer uses bellows in his melting furnaces. He uses 3 cylinders where he compresses the air and drives it into the furnaces. However, this method produces small gusts of wind and does not give the optimal effect. He has improved the method by mounting an air vessel which receives the compressed air, and the outgoing wind blows evenly.

128 For an example of a thunder-ship, a variation on the thunder-house, see Turner 1973, p. 338.

129 The natural philosopher John Canton (1718–1772) invented a strongly phosphorescent compound, 'Canton's Phosphorus', made of sulphur and calcined oyster shells (CaS). It is not clear what kind of objects Nairne showed to Bugge. Perhaps it related to comparison of the aurora/electric light to the intensity of phosphorus.

130 The furnaces of the celebrated ironmaster John Wilkinson (1728–1808) were in northwest England, several days of coach travel away from London. There is no evidence that Bugge met, let alone visited him. He may have heard this information from the cannon founder Verbruggen at Woolwich.

I went to Richmond where I had the opportunity to make the acquaintance of Doctor Demainbray[131] and his son-in-law, Mr Rigaud.[132] They were both very courteous men, and they were kind enough to take me up to the observatory which the King has erected for his own pleasure.[133] In the basement or cellars are mathematical workshops. On the first floor there are several rooms. a.) one housing the transit instrument;[134] the telescope is 5 feet, the axis 3 ½. At the top of the solid block A was a device for the lantern. a and b are two fixed brass sticks with holes, through which the triangle cde can be moved. At the end F the lantern illuminating the filaments is placed. At the other end is placed a bar cg with a counterweight g, so that the lantern remains in the required position.[135] b.) a declination compass. The magnetic meridian is very seldom parallel to the axis of the compass needle, but most often deviates to one side.

September 1777
Richmond

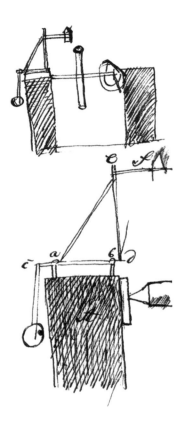

131 Bugge calls him 'Doctor (de) Membray' or 'D.M.' throughout. Stephen Charles Triboudet Demainbray (1710–1782) was first an itinerant lecturer of natural philosophy, and in 1768 he was appointed superintendent of the King's new observatory at Richmond. On his career, see Morton and Wess 1993, pp. 89-119.

132 Stephen Rigaud (d. 1814), observer at the King's observatory at Richmond, was married to Demainbray's daughter, Mary. Their son **Stephen Peter Rigaud** (1774–1839) was to become professor of astronomy at Oxford. No relation to the painter Rigaud whom Bugge had met the previous week.

133 We found no modern study of Richmond Observatory to complement the limited information on the 18th-century period in Rigaud 1882 and Scott 1885, but see Donnelly 1973, pp. 42–44 and the list of instruments in Howse 1986, p. 80, and Howse 1994, p. 217.

134   Howse 1986 lists "1769 / transit instrument / F=8(?) ft, axis = 3 ½ ft / [made by] Adams, London", not known to survive. The next month, Bugge would visit Adams where he "saw a new transit instrument which he is making for the King's Observatory at Richmond". It is not known whether this was intended to replace the earlier transit instrument, or was ever delivered. Millburn 2000 only mentions (p. 105) that the transit instrument at Richmond was seen by Lalande in 1792 and attributed to Adams, but that this cannot be checked as the instrument does not survive.

135 In Oxford, Bugge was to see a superior system to illuminate the wires in transit instruments, which he had his instrument-maker Johannes Ahl copy for the transit instrument made for the Copenhagen Observatory. Pedersen 2001, pp. 36-42.

September 1777
Richmond

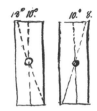

When using the large 12-foot
sector which will be described
later, the observer takes up a
peculiar position. He lies down
on his back on a kind of short
settee where he can raise or
lower the pillow at will. I do
not think that the comfort thus
achieved will be very great.

This is found by making the pin so that the needle may be turned upside down (however, the north pole must remain at the same side). The first time the needle will show a magnetic variation of for example 12°, and the second time a variation of 8°, and the true declination will be their average. c.) The inclination compass was also constructed in such a way that it could be revolved, as it must have the same inclination when it returns to the magnetic meridian. d.) A dividing disc or instrument invented by a Scottish clock maker, but constructed and improved by Demainbray. He stated that it was possible to divide a circle with a radius of up to 3 ½ feet very accurately, although the radius of the actual disk was only ½ foot. D.M. stated that the division was made

very quickly and accurately. In this room there were also several other instruments relating to physics.

The large mural quadrant[136] of about 140° is of very good workmanship. D.M. told me that the Royal Society[137] had had great difficulty to find out how to suspend it, but that he had done it very simply, by taking the Centrum gravitatis of the quadrant and the Centrum gravitatis of the sector separately and suspending them from these two points.

In order to strengthen the quadrant, not only the usual transverse rulers had been placed behind it, but also others parallel to those in front, so that the whole quadrant was double. Demainbray thought that the instrument had thus been very much strengthened. The quadrant was not divided into 96°.

In the same room there was also a small movable azimuth quadrant of about 1 foot radius.

In a room between the transit instrument and the mural quadrant there was an

Eight-foot mural quadrant, radius 244 cm, from the King's Observatory. Made by J. Sisson, London, in 1770. Science Museum London. •
Mural quadrants, fixed to a wall in contrast to portable quadrants which rested on tripods, allowed precise angular measurements to be taken of celestial bodies. As it extends beyond the normal 90° it can also be named an arc. The extension to 142° allowed all the circumpolar stars to be viewed from London using one instrument. ▼

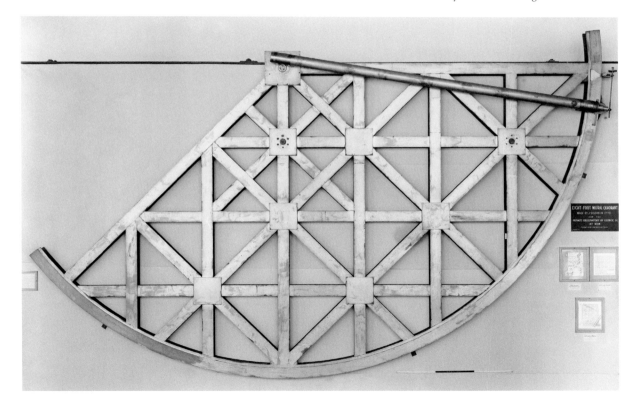

136   Illustrated above. Howse 1986 has this as "1770 / mural arc / 140°,
       R = 8ft / [made by] Sisson, London / now Science Museum inv. 1889-39",
       with reference to King 1954, plate 54.

137   Our interpretation of Bugge's 'vid. Soc', which stands for 'scientific society'.

September 1777
Richmond

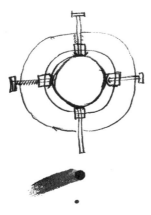

instrument for finding the equal altitudes, according to the description in Smith's Opticks.[138]

Close to the quadrant is a sector of 12 feet radius, with an achromatic objective.[139] It is of excellent workmanship and has all requisite movements. At the top of the objective the verification is made by a triangle movable in a circle. If I remember rightly, Smith has the same idea. It can be rotated to the east and to the west by a wheel placed on top.

At the very top there was a dome, in which a sector of [*blank*] feet diameter was placed on a parallactic support. The roof could be rotated by a very expensive device, consisting of a toothed iron bar which goes round the whole circle, and which can be moved by a wheel and a handle.

Astronomical clocks.[140]

[fol. 43 recto is blank]
[fol. 43 verso contains a list of expenses, among others for four instruments]

The Observatory at Richmond, also known as Kew Observatory. Engraving by G.E. Papendieck, c. 1820. King George III had a strong scientific interest, and had this building erected to enable observation of the Venus transit in 1769. ▶

138   See 25 recto. Not listed by Howse.

139   Howse 1986 has this as "c. 1769 / zenith sector / R=12ft, O=4 ¼in / ? [made by] Sisson, London / to Armagh Observatory".

140   Howse, 1986 and 1994, lists a clock and a journeyman clock, both c. 1769 and by Shelton, London, both now at Armagh Observatory.

To provide a firm and stable site for the transit instrument and the mural quadrant the foundations were laid 20 to 30 feet below the ground.

In addition to these astronomical instruments, this royal building houses an immense collection of all physical instruments. The first beginnings of it belonged to Doctor Demainbray who donated them to the King. The collection increases every day through the instrument makers and artists patronized by the King.[141] Here are also several beautiful objects relating to natural history. The observatory stands in the middle of a beautiful plain. However, it has not a very free horizon. Doctor Membray claimed that the weather was much clearer there than at Greenwich.

September 1777
Richmond

I visited the other Mr Dollond, who lives on the Haymarket.[142] I bought the new parallel ruler invented by him.[143] Neither of the Dollond brothers seems to have any theoretical knowledge.

I saw the Repository of the Royal Society for the Encouragement of Arts, Manufactures and Commerce.[144] I made the acquaintance of Mr Bailey, and for 3 Guineas I bought the first part of his description of machines and subscribed to the second part, which he promised would be

The 24th

◀ *Rolling parallel ruler, signed 'A.G. Eckhardt Junr. Invt . 1770 / Dollond Fecit', width 347 mm. Museum of the History of Science, Oxford. Bugge bought such a drawing instrument from Dollond.*

141 For details see Morton and Wess 1993.

142 This would have been John Dollond junior, see note 114.

143 The new rolling parallel ruler was in fact invented by A.G. Eckhardt, who patented it in 1771 but later transferred the patent to Dollond, see de Clercq 2005b.

144 Bugge called it the London Agricultural Society, but its real name was Royal Society for the Encouragement of Arts, Manufactures and Commerce, now Royal Society of Arts (RSA). Founded in 1754, it was the prime mover in the development of agriculture in England until the formation of the Royal Agricultural Society in 1838; Trueman Wood 1913, p. X. The Repository of Inventions was a collection of mainly agricultural implements, begun in the 1760s, open freely to the public. In the mid-19th century, old models were disposed off, the bulk was presented to the patent historian Bennet Woodcroft, and other items went to the South Kensington Museum; Trueman Wood 1913, pp. 118 and 381. In 1777, the Repository was still in the Society's initial premises in the Strand; later it was transferred to the Society's new premises in the Adelphi, which Bugge would visit on 15 October.

September 1777
London

completed in February and which will costs a guinea, to be paid later.[145]

The collection of machines has many good objects, but also much which cannot be very useful. In general most of these models are of poor workmanship, and are not nice to look at. Among other things there is a clock with only one wheel. It will be described in Bailey, vol. 2.[146]

Mr Bailey told me that the Society is now decreasing. The cause for displeasure and loss of many of the members was a premium on fisheries.[147] They used to have 2000 paying members, and the house could not take them all; now there are only 800 paying members, *besides 300 who have paid 20 Guineas in full settlement and thus are perpetual members*. And there are not many meetings. Nobody is struck off the list who has already paid for one year, unless he expressly demands it.

The 25th Mr Arnold:
Watchmaker. Adelphi.
Timekeeper.

I made the acquaintance of watchmaker Arnold,[148] whose watches are lauded in Captain Phipps's Voyages.[149] He claimed to be able to make such excellent watches with seconds that even after a very long time they would not differ noticeably. His improvements

145 Alexander Mabyn Bailey in 1773 had succeeded his father William Bailey as Registrar of the Society. Bailey 1772; the second volume, published in 1779, duly records in the list of subscribers: "Bugge, Mr. Thomas, Professor of Astronomy, Copenhagen".

146 Bailey, vol. 2, pp. 63–68: Description of Mr Hill's Time-piece, with three plates. In 1773, Matthew Hill was awarded fifty guineas for his clock, which was judged "intirely new, and ingenious in its construction, as it consists of much fewer parts than in common use". Probably = 'Matt Hil, c. 1790; Devonshire Str.' in the List of Former Clock and Watch Makers in Britten 1982.

147 In the 1760s John Blake (d. 1790), a former East India Company captain, set up an organisation to get fresh fish to London. It involved the design of suitable vehicles, for which several wheelwrights and inventors sent proposals to the Society; see Stern 1969–70. The Society's financial involvement in this scheme antagonized many members and contributed to the sharp decline in membership; see Allan and Abbott 1992, p. 104 and Appendix 2.

148 John Arnold (1735/6–1799), watch and chronometer maker, set up business at 2, Adam Street, Adelphi in 1771. See Mercer 1972.

149 Constantine John Phipps (1744–1792), naval officer and politician. Phipps 1774 contains a report (quoted in Mercer 1972, pp. 26-28) on Arnold's pocket marine timekeeper with pivoted detent, which was tried on this voyage.

should be as follows: a.) he rejects spiral springs for adjustment of the watches, because they do not keep their spiral form during their movements. Instead he uses helical springs for his balance. b.) he showed me a spiral spring made of steel and brass and with a long pointer at one end.[150] It can then be seen that the pointer ab makes noticeable deflections as the spring gets cold or warm. c.) the balance for his sea clocks looks approximately as shown in the following figure.[151] The helical spring is placed in A perpendicular to the plane of the balance. ab and cd are two pieces of brass with screw threads in b and c for adjustment. The rest is of steel. He claimed that when ab expands with heat or contracts with cold, the vibrations will also be changed and adjusted. d.) Finally, he has also improved the pendulum of the astronomical clocks, which he promised to show me on Monday 29th together with several relevant experiments.

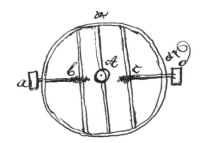

In his workshop he had a transit instrument, which belonged to Graham.[152] He tests his watches against this instrument.

150 This was not actually a spring at all, but a compensation device, taken directly from the design by Thomas Mudge known today as a 'chelsea bun' compensator. We thank Jonathan Betts for this comment.

151 This can be recognized as the so-called Double T-Balance, of which Arnold made some twenty in the years 1778–1780. See Mercer 1972, chapter V: John Arnold's Balances.

152 George Graham (c.1673–1751), horologist and maker of scientific instruments. In 1713 he succeeded his erstwhile partner in business, Thomas Tompion, to become London's most skilled and influential maker.

September 1777
London

Catal.

An astronomical clock or regulator = 90 Guin., but then all holes are made in diamonds, yes even the hooks on the forks that go into the escape wheel.

A pocket watch with Seconds, in Gold, 80 Guin.

The same, in Silver, 60 Guin.

Mr Magelhaens [153] was kind enough to pay me a visit. He believed that Arnold was able to make good astronomical clocks, but that most of his projects concerning sea clocks were charlatanry.

At the Black Friars Bridge there are 4 glassworks. I saw one of them, but found nothing remarkable.

In the same place, at Jacob and Viny,[154] there is an installation for making wheels in a new way; he bends the wood, which makes up the entire rim of the wheel. Thus all the layers of the wood will be parallel to the surface B. This gives a greater strength than when they are cut in the usual way (A).

Neither does he drive in the iron hubs with rivets, but they are screwed in with screws.

▲ *Dial of barograph clock for continuous recording of atmospheric pressure, made by Alexander Cumming in 1766. Science Museum, London. This is the clock that Bugge saw in Cumming's workshop.*

A set of his wheels (2 front wheels and 2 back wheels) costs 8 Guineas, whereas the usual ones cost 6 Guineas. Of course, these oak wheels last much longer than the ordinary wheels. His entire plant and wood store, as well as the curved rims for wheels, the wares and the spokes, were large and neat.

The sculptor Charles Harris,[155] Statuary, opposite to the New Church in the Strand, has very beautiful statues, vases, bas-reliefs of gypsum, marble, and a great variety of English stones. He gave me a catalogue.

Catal

I went to see Mr Cumming, a very competent watchmaker *A. Cumming, Watchmaker, New Bond Street.* [156] I bought his work, The Elements of Clock and Watch Work; by Alexander Cumming, 18 sh.

As regards watchmaking he showed me: a.) A Barometer Clock[157] which plots the barometric height; abcdefgh is a disk of parchment which remains on the watch for 3 years (in order to avoid the leap year). The inner circle has been divided into 365 parts or days, and these have again been properly divided into months, January, February, March, etc.

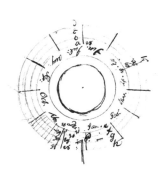

153   The Portuguese Jean-Hyacinthe (de) Magellan (1722–1790) lived in London from 1764 onward and was elected a Fellow of the Royal Society in 1774. He designed instruments, published on them and acted as consultant and middleman in supplying instruments all over Europe.

154   *The London Directory for the Year 1778* lists (p. 90): Jacob and Viney, coach wheel makers, St. George's Road, Blackfriars Bridge. The Benjamin Franklin Papers (http://franklinpapers.org/franklin) contain correspondence (1783–84) with the wheelwright John Viny which shows that he worked with Franklin to develop a new method of making wheels in one piece; that Viny and Jacob quarreled over the patent rights after Jacob patented the method in 1771; and that Viny continued to operate the firm for the creditors after it went bankrupt in 1778.

155   Charles Harris (d. 1795?), Strand, London, sculpted monuments in the best 18th-century tradition, with a lavish use of coloured marbles and reliefs, and obtained commissions from families who patronized leading sculptors; Gunnis 1968, pp. 188–190 and Whinney 1988, pp. 273–274.

156   The watchmaker and mechanician Alexander Cumming (1731/2–1814) established himself in New Bond Street in 1763.

157   In 1765, Cumming made an eight-foot tall barograph clock for King George III, the first effective recording barograph. The following year, he made a less ornate one for his own use, which Bugge saw on this occasion and again a month later. After Cumming's death it was bought by the meteorologist Luke Howard, who used it for the observations that formed the basis of his pioneering work *The Climate of London* (1820). It is now in the Science Museum, London and illustrated opposite. Three more, later barograph clocks by Cumming are known, see Collins (2002).

September 1777
London

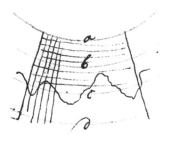

The first circle a is 28" of the barometric height, the second b is 29", the third c is 30", the fourth d is 31". P is a pin with a pencil which is carried up and down by the barometer according to its height, and at the same time it writes on the parchment. When the disk is rotating for a whole year by means of the clock mechanism, then a correct, unbroken curve of the barometric height will be traced on the parchment.

Mr Cumming has drawn the inches so that 31" is represented by the outer circle and 28" by the inner circle, but he has done so solely in order that the inner part of the dial could represent the earth, and the outer part the atmosphere.

It is easy to visualize that each curve ab, bc, and cd is divided into concentric circles, and that the radius or the cross line for the division of each day is traced. b.) he showed me more exactly the composition of the gridiron or the ordinary compound pendulum which is supposed to be John Harrison's invention. c.) Ellicott's pendulum. d.) the same, improved by Cumming. Furthermore, everything has been thoroughly described in his above-mentioned work.[158]

158 Cumming 1766.

On the top of his house he has a nice small observatory.[159] For making clocks he has a very beautiful transit instrument made by Ramsden.[160] The axis is about 2 feet. And the telescope is achromatic with 3 1-foot objectives. The telescope was fitted with several considerable improvements:

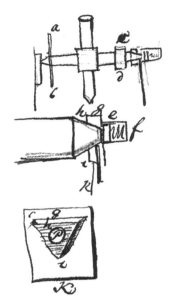

a.) At one end a declination circle ab was placed; and at the other end a counterweight cd.

b.) At one end was placed a small spring ef in order to create some resistance when the instrument expands with heat.

c.) Above the bearings in which the pivots are moving, a cover ghik is placed, in order to avoid that the pivots and the devices for changing these get covered in dust.

It can be observed that the front plate has been cut through at P, so that the movement of the conical pivot is completely free.

d.) The heads of the screws moving the bearings had been divided, so that it is

159 This observatory is not listed in Howse 1986 and 1994.

160 Bugge would visit him on 30 September. On Ramsden, see McConnell 2007.

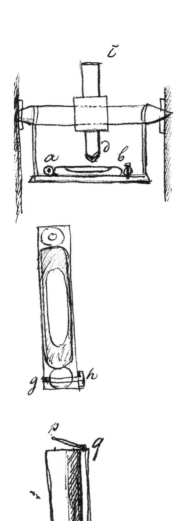

possible to determine how much the instrument is corrected each time.

e.) When the spirit level ab had been adjusted by proper reversal in the horizontal position of the meridian circle cd, then Mr Cumming noticed that the bubble did not remain in the centre when the telescope was elevated to 20°, 30°, 40°, etc. From this he concluded that the axis was not on the same level or on a level parallel with the length of the spirit level. In order to change this he had fitted a screw gh, so that the entire spirit level can be moved to both sides.

The adjustment of this screw must then be continued until the bubble remains at its marks in all positions of the meridian circle.

f.) In order to open the lid pq a rod qr goes down through the central cube towards the eyepiece, so that the lid can be opened by pulling it by hand.

Mr Cumming showed me a very beautiful equatorial instrument by Ramsden. It has not been fitted up according to the figure in Martin's Philosoph. Brittan.,[161] Vol. 3, p. 404, but it looks most of all like that of Dollond in his Description and Use of the Equatorial Instrument.[162] It has the advantage that the Centrum gravitatis of the whole instrument is always steady and supported. This can be verified by reversal. However, it must be pointed out that Ramsden has mounted a spirit level below the tube which is hanging in pivots, i.e. always vertically in any position of the telescope.[163]

The 29th I went to the watchmaker Arnold in order to see, at his invitation, some experiments with his compound pendulum for correcting for heat and cold. His pyrometer was mounted with wheelwork so that the smaller circle A was divided into 14 parts; its pointer is ab; while ab goes one part = 1/14 the

September 1777
London

The 29th Septemb.

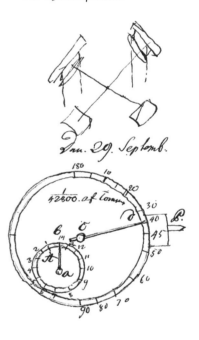

◀ *Portable equatorial instrument by Ramsden, London, made between 1775 and 1789. Overall height with telescope horizontal 73 cm. Teyler's Museum, Haarlem. Cumming showed Bugge a similar instrument.*

161 Martin 1771.

162 See the plate in Dollond 1771.

163 Ramsden 1774, pp. 3–4: "There is a small Brass Rod (M) placed immediately under the Telescope (…) On this brass rod there is occasionally placed a hanging Level (N) (…)".

London 1777
September the 29ᵗʰ

pointer cd completes its entire revolution and divides 1/14 into 180 parts. It was so constructed that 1/180 of the outer circle was = 1/42800 of an English inch.

The total circumference of the large circle = 1/14 of the small = 1/238", and the entire small circle = 1/17" = 0.0588".

At one side a small point P sticks out; which gives the wheel-work its total movement. At the slightest pressure the pointers are running.

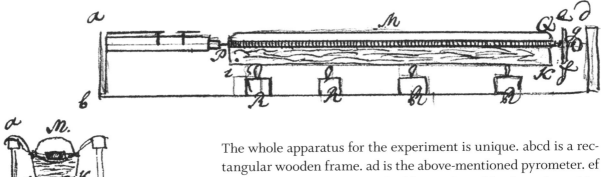

The whole apparatus for the experiment is unique. abcd is a rectangular wooden frame. ad is the above-mentioned pyrometer. ef is a device which can be pushed forwards and backwards; and by a small screw at g the whole pendulum PQ can be adjusted to the point P; and the pyrometer pointers can be set to 180°; PiK is a rectangular tin box which is filled with water and covered by the cover M. The lamps are placed at R.R.; Arnold made his first experiment with an ordinary rod. He filled the box with water and placed three lamps below it. Then it could be observed that the pointers moved quite steadily until the water was boiling; then they stopped. The rod was made of a metal invented by Arnold himself. Its hardness and expansion was something between steel and brass. It expanded 4 ½ times.[164]

He made his second experiment with a ½ second pendulum. Here the pointer showed no movement except that the long pointer moved 10/180, which Arnold told to be on purpose in order to counterbalance the greater or smaller density of the air from summer to winter.

He made his third experiment with a 1 second pendulum, which was so constructed that it contracted with heat. By cooling it he showed that when he poured water over the central bars,

164 Presumably Bugge means that the co-efficient of expansion of the metal was 4 ½ times that of steel.

London 1777
September
the 29th

the pointers took one direction, and when he poured it over the side bars they took another direction.

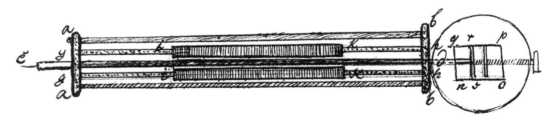

Arnold's pendulum consists of 5 rods. The outer two ab and ab and the central cd are of steel. The two others gh are compounded, first of brass gi, then of Arnold's metal ik, and finally of brass kh. *The metal ik contains much zinc.* Finally qnop has been cut out on the very lenticular bob. The bob is hanging on the plate rs, which Arnold has placed between the Centrum Oscillationis of the pendulum and the Centrum Gravitatis of the bob.

I visited the well-known portrait painter Sir Joshua Reynolds.[165] His portraits were numerous and most excellent. The postures of the persons portrayed were well chosen and characteristic. For a life-size portrait he charges 150 Guineas, for a half-length 90 G. and for a portrait without hands 35 G.

165 Bugge wrote "Chevalier Regnald". Sir Joshua Reynolds (1723–1792), knighted 1769, first president of the Royal Academy, was the leading London portrait painter since the 1750s.

I visited Master Sisson, living at the corner of Beaufort Buildings in the Strand.[166] First he showed me a model of a private observatory fitted up by Mr Nathaniel Pigott Esq., Framton House near Cowbridge, Glamorganshire in Wales.[167]

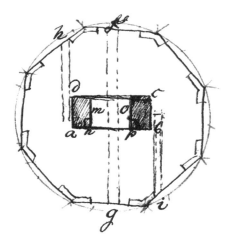

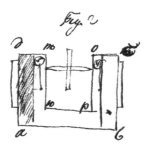

It is octagonal. abcd is a square stone in one piece, the shape of which is illustrated by the profile in Fig. 2, abcd mnpo. The space mo is for the transit instrument whose opening is f.g. In n and p corners have been carved right down to the floor; there the clocks will be hung up. ad and bc are for two mural quadrants, one towards the south, the other towards the north, and with their necessary openings in the ceiling ah and bi. Towards the south a small equal altitude instrument will be placed in a window.

166 Jeremiah Sisson (1720–1783/4), maker of mathematical instruments like his father Jonathan Sisson. When his father died in June 1747, Jeremiah took over a thriving business and the premises at the corner of Beaufort Buildings, Strand.

167 The astronomer Nathaniel Pigott (1725–1804) erected an observatory at a house he rented at Frampton near Llantwit Major, Glamorganshire, Wales, with a transit by Sisson, a 6-foot Dollond achromatic and several smaller telescopes. In 1781 Nathaniel and his son Edward moved to York, where they built a substantial two-storey stone observatory in the garden to house the instruments from Frampton. See McConnell and Brech 1999. Howse, 1986 and 1994, does not list this observatory, but does list (1986, p. 83) Pigott's observatory in York.

Sisson showed me a transit instrument made for this observatory. The axis and the telescope were about 2 ½ feet.

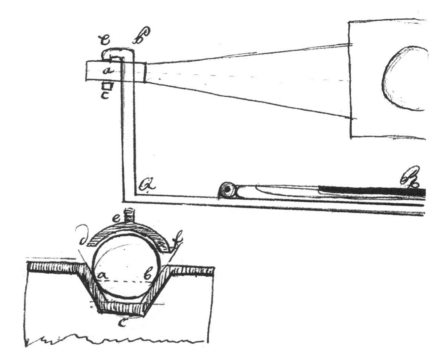

The axis of the transit instrument in its bearings. a and b are the centres of suspension.

Sisson made his pivots or the axis on which the instrument rests both cylindrical and of exactly the same diameter by letting them slide through a hollow test cylinder. He has another excellent arrangement, namely that the centre of suspension P of the spirit level PQR must not be in P but in e, just above the centres of suspension a and b of the whole instrument. Furthermore, all screws should be made of brass. If they were made of steel there would be a difference in the expansion which would twist the instrument.

As the sight is indistinct and also somewhat uncertain when you have to watch through edges, Mr Maskelyne has made a device enabling him to move the eye-piece sideways.[168] In the two segments acb and def, placed at the end of the telescope acbefd, dovetails have been cut out, in which the piece ghki can move. Attached to this is: 1.) The tube of the eyepiece P 2.) lm with teeth which can be moved to both sides by the toothed wheel Sfp.[169]

As regards the Linea Fiduciæ,[170] it makes no difference. When the picture CD has been determined in the right place on the filaments without Parallaxis, then the picture will be visible

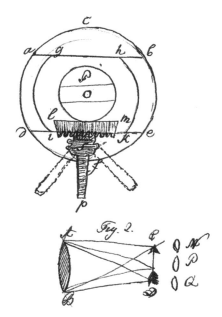

168 Howse 1975, pp. 35–37, describing Bradley's 8-foot transit instrument, mentions the "single positive-focus eyepiece capable of being moved laterally to view each of the five wires separately", and discusses how this was used in the determination of the Greenwich meridian from 1750 to 1816. Bugge later reports that professor Hornsby in Oxford was not enthusiastic about this invention by the Astronomer Royal.

169 Bugge seems to have written Sfq or Sfg, but the drawing suggests he means Sfp.

170 Presumably, Bugge uses Linea Fiduciae, Fiducial Line, as another term for alidade or diopter.

September 1777
London
Mr Ramsden Piccadilly

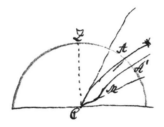

at the eyepiece either in P or in Q or in N.

I visited Mr Ramsden who at once showed me an 8-foot mural quadrant which was in preparation.[171] He had placed it on a very large table, standing immediately on the pavement. He suspends it from two points.

For Greenwich and Oxford he is to make equatorial instruments with an axis of 9 feet. The diameter of the matching circles, the horizon, the equator, etc., will be 6 feet.

Ramsden's ideas about a new instrument.

When a ray, coming from a star, passes through the atmosphere, it will be bent towards the zenith, as long as the density of the air increases constantly from A to C. But if, however, the air is thinner in certain strata from M to C, then the concavity will decrease towards Z. And this would necessarily cause a great change in the quantity of the refraction. Ramsden claimed that the more Phlogiston there was in the air, the more variable the refraction would be; and that this had no connection with the barometer

171 This was the quadrant for Padua University, now in the Museo La Specola, INAF-Astronomical Observatory of Padua, Italy; see McConnell 2007, pp. 93–97.

or the thermometer. His design for a manometer indicating the change of refraction is as follows:

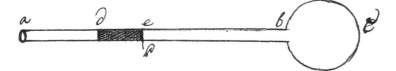

abc is a tube and a sphere, both with a known diameter and content. de is a column of mercury filled into the tube.

For the rest ebc had been filled with ordinary air. When the barometer is 30 English inches, and the Fahrenheit thermometer 55 (which is the state of the air for most refraction tables) the point P is marked on the tube, and the rest can be divided into thousand parts by means of a rule. When the elasticity of the air outside increases, the mercury will come nearer to c, and vice versa.

As it is difficult to keep the tube in a horizontal position, he has wound it round a cylinder, which after all gives the same results. It is easy to understand the whole thing from Fig. A and B; the latter corresponds to what has been mentioned above.

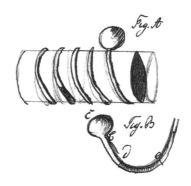

◀ *Jesse Ramsden (1735-1800) by Robert Home, c. 1790. Oil on canvas. The Royal Society, London. Ramsden is portrayed with his dividing engine. In the background the enormous astronomical circle that he made for Palermo Observatory.*

Ramsden pointed out quite rightly, that the objective micrometers had the following shortcomings: 1°) it is hardly possible to grind the cut glasses for the micrometer of gd. The centre C and the periphery a and d and c and b have different focal lengths or focal points. 2°) even when achromatic glass is used, some colour or irradiation remains, which will have the effect that the diameter of the planet is found too great.

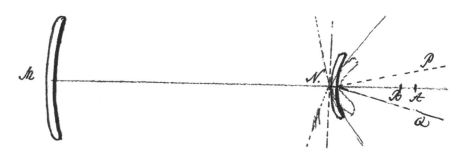

Let M and N be the objective and the ocular mirror in a Cassegrain telescope. A is the Centrum Curvaturae, and B is chosen as the Centrum Motus for the mirror N (cut at NB). When the mirror is moved, it will be possible to measure the diameter. For the angles of the radii will be the same as the angles of the tangents.

John Bird (1709-1776). Mezzotint by Valentine Green after C. Lewis, published 1776. On the table a drawing of a mural quadrant, with a beam compass lying on it. ▶

As regards the spirit level of Sisson's transit instrument it must be observed that the plate or the glass of the spirit level has been marked with points at the end of the bubble, to observe its rise and fall or the inclination of the plane.

The second October[172] I was introduced by Dr. Solander[173] into the Royal Society Club[174] in the Mitre Tavern. The Society has dinner there every Thursday, whereas a non-member is allowed to come only every fortnight. The president, Sir John Pringle,[175] invited me to dinner the following day.

I visited Mr Russel, Strand, near Somerset House, Nº 156.[176] He showed me a small scale rule by Bird,[177] by which an inch on the transversal lines was divided into 500 parts. AB was divided into 10 parts, each part being so large that it could be subdivided into 5 parts as shown between B and C.

Furthermore, Russel had a 2-foot rule

The 5th October. Mr Russel

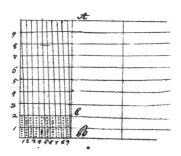

172   Bugge wrote September, clearly a slip.

173   The Swedish botanist Daniel Charles Solander (1736–82) was since 1773 Keeper of the Natural History Department of the British Museum.

174   Dating back to about 1731, its history has been documented in Geikie 1917.

175   The physician Sir John Pringle (1707–82) was president of the Royal Society from 1772 to 1778.

176   Although Bugge never mentions his first name on any of their three encounters, this was William Russell (?–1787), FRS, whom Jean Bernoulli had visited in 1769; and not (as suggested in Pedersen 2001, pp. 42–43) the portrait painter and astronomer John Russell, famous for his lunar mapping. The listing of his private observatory in Howse 1986, p. 77 is based on Bugge's journal, which Howse already knew at the time. On Russell, see de Clercq 2007.

177   John Bird (1709–1776) came to London from the county Durham. He was employed in the workshops of Jonathan Sisson and George Graham, before setting up his own workshop in York Buildings in the Strand. He was one of the greatest of London's makers and supplied several astronomical observatories with their main instruments.

by Bird with a Nonius or a Vernier by which inches are divided into 1000 parts. The construction of this instrument is as follows:

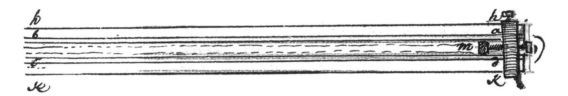

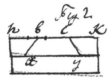

The central rod abcd can be moved forwards and backwards by the dovetail shown in profile in Fig. 2, bcyx. The fixed ruler ahbh is divided into inches, and the movable ruler has a Vernier division. hk can slide along the two rulers, and it can be fastened by the screw h. This gives a rough adjustment, whereas the free motions are made with the screw Smp.[178] With this rule I compared an inch rule, bought at Smith's, and I found that 24 inches were 1/10 inches too short.

Mr Russel also had a very beautiful transit instrument by Bird. The axis

---

178 This is puzzling: only 'm' can be seen in the drawing.

was 2 feet; the telescope was 4 feet. Bird considered the spirit level to be the best he had ever made. I also saw one of Graham's astronomical clocks with a gridiron pendulum; it had a peculiar device for correcting the pendulum if the brass and steel rods did not have the correct ratio, or if the pendulum changed with warm and cold.

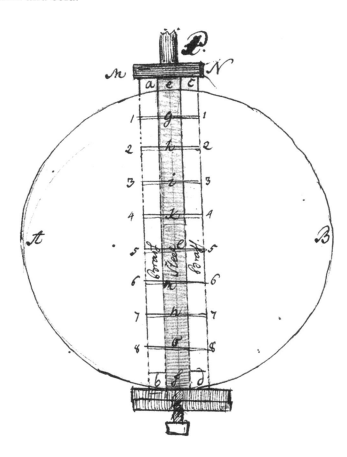

October 1777
London
The 5th October

0,25" 0,56"

AB is the lenticular bob of about 6 inches diameter. The steel pendulum rod P has been fastened by screws in a round brass piece MN from which goes out a brass crutch consisting of two rectangular pieces ab and cd. Between these is a rectangular piece of steel ef which supports the lenticular bob. The brass and the steel rods have both been pierced at certain assumed intervals 1.1., 2.2., 3.3., etc.

If now the compound pendulum (gridiron) had too much brass and too little iron in it, or if it expanded with heat, *then a pin is put through for example 1g1* and then the brass ae and the steel gf have the effect that the brass 1b and 1d is eliminated as well as the steel eg. Consequently the steel rods get longer, and the expansion becomes just as much smaller.

If the pin is put through 2h2, then only the brass a2 and c2 is active, whereas 2d and 2b is eliminated. The steel hf is active, but it is eliminated, and consequently the steel of the gridiron gets longer by a small piece.

If, on the contrary, the compound pendulum contains too much steel and too little brass, then the pin is placed in some of the holes below the centre, for example in 7n7. Then of the brass a7 and c7 are active, but of the steel only nf.

It may be objected that for a complete elimination of expansion and contraction 9 rods are necessary, as the ratio of expansion between brass and steel is 5:9.

[From here to the end of , fol. 57 verso, Bugge wrote down lists of books and instruments bought in London; for the transcript and comment, see the appendices].

Doct. Radcliffs Library

I Oxfort observerer jeg en stor staaende
mit paa Gaden og byggaet i en Cirkel.
Den nederste Deel deraf er aldeles
uden til nogen brug, men er som en
slags promenade. Den anden Etage
er et Bibliothek. af Radcliff, hvilket jeg
sig er meget stoort.
finde trende Inguer som man kalte den
Lyksalige, i et orguerne at Hadriani
Imperatoris Villa og givet til Oxfort
Academie at Newdgate 1775.

g 7 den tredie Etage
er og en Samling at
bøgst. og fra Domen
ualte Latmen.
bygnugr Ashe er
efter en meget god
smag; og Ornamen
ter er engang ved
bragt.

Jeg besøgte hr. Hornsby i sit nye Observatorium, hvilket han med megen be-
enduellighed viste mig. Eftersom et
gerut udkast til Tanken der til blive
følgende

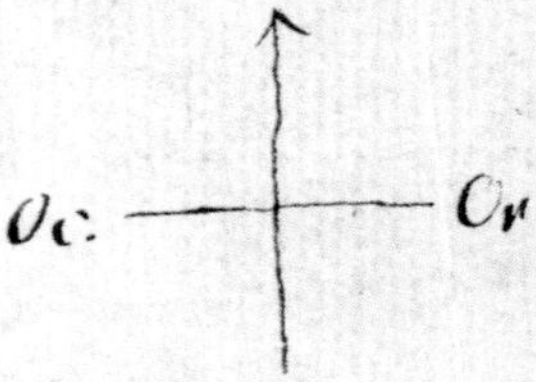

A. er Mr. Hornsbÿs Huus; B; er en Gang
fra hans Huus til Observatorium. Men
egentligen det virke Observatorium bestoer
i trende Deele. Nr 1.) for bygga Murad.

Nr A

In Oxford I saw Doctor Radcliffe's Library, a circular building, standing in the middle of the street.[179] Its lowest part is not used but serves as a kind of promenade. On the second floor is Radcliffe's small library in which there are two figures (called candleholders), dug up in Hadriani Imperatori Villa and given to the Oxford Academy by Newd[i]gate in 1775.[180] *On the third floor there is a collection of books, and at the top the dome or lantern. The style of building is in good taste, and nowhere did they save on ornaments.*

I visited Mr Hornsby[181] in his new Observatory,[182] which he showed me with great willingness. From a sketch the idea will be as follows.

A. is Mr Hornsby's house, B. is a corridor from his house to the observatory. DC is the real observatory divided into two rooms. 1.) for both mural

October 1777
Oxford
The 6th

View from the South

179 Dr John Radcliffe (1652–1714) left £140,000 in his will for the enlargement of University College, for travelling medical scholarships and for a library. This library, the Radcliffe Camera, was completed in 1749. See Guest 1991.

180 Sir Roger Newdigate, fifth baronet (1719–1806), politician and architect, an early exponent of the Gothic revival in architecture. During a tour through Italy in 1774–76, he amassed a large collection of paintings, statues and other artefacts, including two marble candelabra from Hadrian's villa at Tivoli which he donated to Oxford University. They are now in the Ashmolean Museum.

181 Thomas Hornsby (1733–1810) succeeded James Bradley as Savilian professor of astronomy in 1763, when he also became professor of experimental philosophy. See Wallis 2000.

182 Although Dr Radcliffe's will made no mention of it, an observatory was also financed from the residue of his benefaction on the suggestion of its trustees. This Radcliffe Observatory was the second permanent British observatory after Greenwich. Building began in the early 1770s but was not completed until the 1790s. For a discussion of its erection and instrumentation in the Hornsby era, see Guest 1991, pp. 224–251 and Bennett 1993.

October 1777
Oxford

quadrants. 2.) for the sector. 3.) for the transit instrument. In the wing FE[183] smaller instruments will be set up, for the use and practice of young students.

The tall building or tower GHI is placed in the middle. The central floor EDIH will be fitted up as a reading room and with two or three smaller rooms. The telescopes are kept on the second floor, from which it is possible to survey the whole horizon.

Everything is very nicely decorated, and is in a good taste. The building will probably be finished within three years.

The 7th

I am now going to describe each room separately.[184] In No. 1 there are two mural quadrants with a radius of 8 feet, placed on the same wall. One faces south and the other faces north. They are both made by Bird as described in his Construction of Mural Quadrants.[185] The telescopes are achromatic and have been improved by means of a small weight, in order to prevent friction of the telescope around its centre. The vertical was indicated by a fine silver thread.

183 Bugge erroneously wrote FD.

184 See the list of Radcliffe Observatory instruments in Howse 1994, 217, replacing the entry in Howse 1986, 79. Most of these are now in the Museum of the History of Science, Oxford.

185 The 1768 treatise that Bugge bought in London.

In the same room there were also two 3½-foot Dollond telescopes with three glasses. One was mounted on a kind of parallactic support. ab is the axis mundi. cm is a circle by means of which the tube could be moved around both in a fine and a rough way.

Furthermore, this telescope was fitted with an object-glass micrometer, newly invented by Dollond.[186] The glasses were as shown in a and b. Hornsby thought that it would be impossible to find such glasses with a radius of 6 inches.

There was also another apparatus constructed according to an idea of Baillie (a Frenchman), by means of which it was possible to reduce or to increase the light during observations of Jupiter's satellites.[187] Two right-angled pieces M & N are moving, one above the other, as shown in Fig. 2. It is then evident that the aperture can be adjusted at will. Mr Hornsby did not consider that to be of any real practical use.

The other 3 ½-foot telescope with three glasses was made of mahogany and was mounted on an ordinary support. This telescope does not belong to the observatory, but to Mr Hornsby himself.

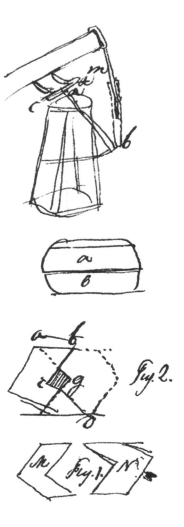

186 See Maskelyne 1771.

187 Jean Sylvain Bailly (1736–1793) published two treatises on the satellites of Jupiter. Bailly 1766 and 1771.

October 1777
Oxford

In No. 1 there were also barometers, thermometers, and instruments for the observation of rainwater. This is done by weighing the water running down into the glass cylinder mn from a funnel of two feet square, which was placed on the roof.

In No. 2 there is a 12-foot sector, constructed according to special principles so that, without doubt, it is the best of all sectors. AB is the axis of mahogany which has been cut into 4 pieces, a,b,c and d. These have again been glued together. Then they have been pressed or tied together by means of the screws p,q,r,s,t,x,y,z. The pivots of the sector are movable in a four-sided frame which, by means of the screws efgh, can be adjusted until the hair is visible on o from both sides. In addition a horizontal

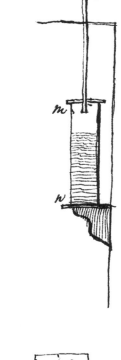

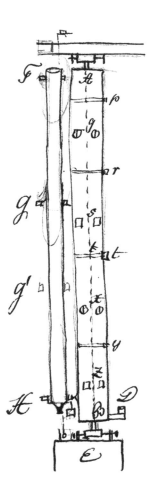

level was placed at D; and it did not move from its marks during the revolution, which shows that the instrument was very well adjusted. E is a stone block placed in the solid gravel ground. This block E is shown in plane in Fig 3. ik and lm are two brass plates attached to the quadrangular piece AB. By means of the screws α, β, γ, δ they can be placed exactly in the meridian. FI, GK, and H are rings in which the achromatic telescope can move. At M the arc HL = 14° is fastened by screws. Its divisions are subdivided by means of two micrometers.

No. 3. Here we find a transit instrument mounted with an 8 foot achromatic telescope. The spirit level was suspended in the same way as shown by Cumming;[188] the bar connecting

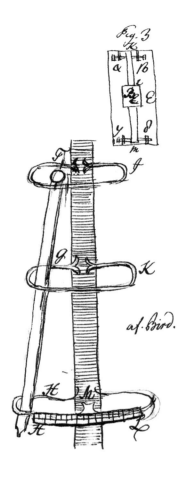

---

188 See fols. 47 verso – 48 recto.

October 1777
Oxford

the yokes was made of mahogany.

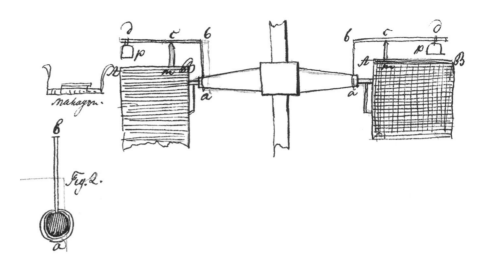

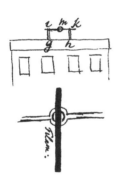

The total weight of the instrument was more than 100 lbs, and in order to relieve the weight from the pivots, levers (bcd) are placed at its outer ends. Their centres of suspension c rest on the iron bars AB. bd and ba are made of mahogany. The latter (ba) is shown in Fig. 2. The weight p counterbalances half of the weight. With one finger I could lift the instrument out of its bearings at a.

The meridian mark ghik has been erected on a house half an English mile away. m is a circle with a hole in its centre through which the sky is visible. The central hair divided this hole exactly.

Two men are required to reverse this heavy instrument, and Mr Hornsby has noticed that if he touches the axis with his hands it expands, so that it was impossible to make it correspond with the meridian marks on both sides.

For the reversal he uses the following support: AB is a round disk standing on four legs tied together at the bottom and in the centre by means of a circular strip. At C the bar DCE is threaded and goes through the hole at C. The bar DE is fitted with a cradle FGHI which, at F and I, can grasp the axis of the transit instrument. When the bar DE is screwed up the whole instrument is lifted out of its bearings. Then the axis alone is turned half-way round so that F and I change places. Then the instrument has been reversed; and when DE is screwed down again, it falls down into its bearings.[189]

The device for illuminating the filaments

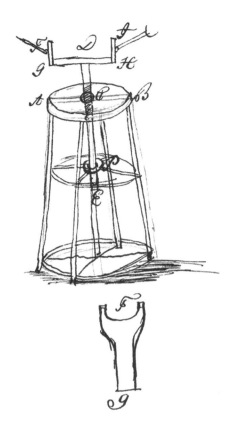

---

189 On 29 January 1771, the instrument-maker John Bird wrote to Hornsby that the transit instrument would "require a weighty Tube, and consequently a very strong Axis, which altogether will be too heavy for two men to turn, end for end, with safety, but I can make an Apparatus with which it may be turned by a Boy of 12 years of age, and without danger"; quoted in Guest 1991, p. 236.

October 1777
Oxford
The 7th

is as follows:

A is the axis of the transit instrument. ab is a circle of mahogany round the axis. It has been attached to the pier without being connected with the axis. Around it is another circle cdef, connected with the piece hg which is of about the same length as the telescope. Q is the lantern, illuminating the circular piece of shining brass MN, with the requisite aperture O. At P a weight counterbalances the lantern. When the telescope has been adjusted to focus on the star, the observer, with one hand, moves the piece hg and the lantern Q

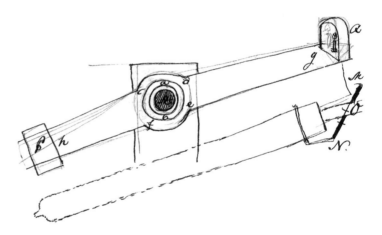

In the Radcliffe Observatory, Bugge described the transit instrument and made a drawing of the system to illuminate the filaments in the objective. The engraving is from Bugge 1784. ▶

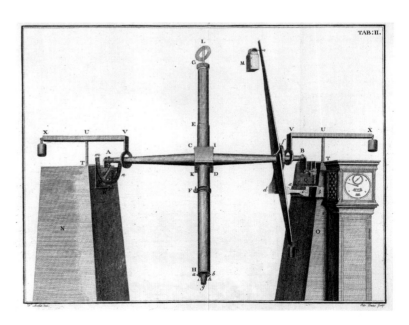

up and down until he finds that the filaments are well illuminated.[190]

In this room there was a clock by Shelton.[191] Mr Hornsby had made it beat the seconds louder by increasing the weight from 9 lbs to about 15 lbs, which he rightly thought to be more accurate than a counter.

On this occasion he remarked that the observatory of Mylord Macclesfield[192] was in possession of a clock with Graham's mercury pendulum which goes just as well as the newer ones with gridiron pendulums.[193] This has been verified by observations and comparisons over several years. This Graham pendulum has not been properly described anywhere; at Mr Arnold's in London I remember having seen Graham's portrait with a rough drawing of the pendulum in the background.[194] According to what Mr Hornsby told me, the arrangement will be as follows: ACB is the entire pendulum. AD is the brass rod. DB is mercury in a glass cylinder which, at C, has been attached to the pendulum rod AC. This rod expands from the

190 As can be seen on the engraving Bugge had his instrument-maker Johannes Ahl copy this lighting device on his own transit instrument, as discussed in Pedersen 2001.

191 John Shelton was a prominent London clockmaker, who among others made five astronomical regulators for the Royal Society for timing the transits of Venus in 1761 and 1769.

192 George Parker, 2nd Earl of Macclesfield (c. 1697–1764), astronomer and politician, president of the Royal Society from 1752 to his death. In about 1739 he erected an astronomical observatory at Shirburn Castle, Oxfordshire; Howse 1986, p. 80. In 1761 Hornsby had observed the Venus transit here and developed his interest in astronomical observations.

193 According to the George Graham entry in the Oxford Dictionary of National Biography, the Earl of Macclesfield possessed a sidereal regulator by Graham in his library, which was fitted with a temperature compensated mercury pendulum (a glass of mercury for a bob) as invented by Graham in the early 1720s. Three weeks later, Bugge was to see the very same regulator during his visit to Alexander Cumming.

194 Graham's portrait, reproduced here, was possibly painted for the second Earl of Macclesfield, who owned it. Presumably, Arnold possessed a copy of the mezzotint made by Thomas Ryley, c. 1750.

▲ *George Graham (c. 1673-1751) by Thomas Hudson, c. 1735-40. Oil on canvas. Science Museum, London. The mercury pendulum can be seen in the background.*

October 1777
Oxford

The 9th

▲ *The astronomical pendulum clock
made by Mudge and Dutton about
1762. The clock is now in Kroppedal
Museum, Høje Taastrup, Denmark.*

centre of suspension A to C; or downwards; and the mercury expands from the bottom B upward to D. Consequently the first expansion should counterbalance the other so that the centrum oscillationis remains unchanged.

Mr Hornsby does not consider Maskelyne's device for moving the eyepieces to be of much use.[195]

Dutton (the maker of the astronomical clock in Copenhagen) constructs sea clocks.[196] Hornsby has tested one of them and found it extremely good. During a transport from London to Oxford and from Oxford to London it correctly indicated the meridian difference of time to be 2" to 3".

Hornsby has determined the Arcturi motus proprius et singularis in the Philosophical Transactions.[197] He uses the Triduum Roemeri[198] and finds the deviations for 70 and 69. As he has found several misprints in it,

195   Cf. fol. 51 recto.

196   William Dutton (1738–1794) was apprenticed to George Graham in 1738 and received his freedom in 1746. He was a liveryman of the Clockmakers Company from 1766–94. He entered into a partnership with Thomas Mudge in 1755 and took over the business in 1771. Edward Legg, The Clock and Watchmakers of Buckinghamshire (webpage).

197   Arcturus is the brightest star in the constellation Bootes, the third brightest star in the night sky, and has the largest "proper motion" – motion across the sky – of any of the bright stars except Alpha Centauri, the nearest star to our solar system. Although others had already detected this motion, Hornsby was the first to measure it, and he presented his findings in Hornsby 1773. In Hornsby 1798, he was to give further results on proper motions.

198   *Triduum Observationum Tusculanorum Roemeri*, Ole Roemer's observations from his Tusculum observatory conducted on 20–23 October 1706, as published by his pupil and successor Peder Horrebow in Horrebow 1735, pp. 157–230. The Kroppedal Museum in Høje Taastrup, outside Copenhagen, is situated near the site of Roemer's observatory, and has a permanent exhibition about Ole Roemer and Danish astronomy.

he gave me a list of them so that I could compare them with the original manuscript in the Round Tower or at the University Library.[199]

The dimensions of the transit instrument are as follows:

The length of the axis	4 feet	2.9	inches English measure
The focal length of the object-glass	8	2.0	
The aperture of the object-glass	-	4.1	
The cube of the axis	-	6.6	
The diameter of the cylindrical pivots in the bearings	-	1.6	
Rad. circul. Declin.	1	5.0	

The dimensions of the large 12-foot sector:

The axis, which has been glued and screwed together out of 4 pieces, is quadratic, each side	-	6	inches English measure
The focal length of the object-glass	12	-	
The aperture of the same	-	3.8	

The arc has been divided into 7 degrees on each side of the zero mark: each degree is divided into 20 parts of 3' each, and these again are subdivided most accurately by means of a wheel micrometer.

199 Bugge returns to this in more detail on fol. 69 recto – 70 recto; see appendix 4.

October 1777
Oxford

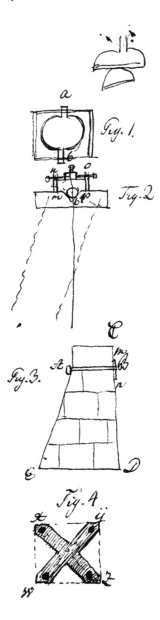

Fig. 1.

Fig. 2

Fig. 3.

Fig. 4.

At the top is a Hooke's joint with two knee joints for the fitting of the axis. To the joint has been fastened a quadrangular frame of 8 English inches square in which the axis ab is placed. Fig. 2 shows this in profile together with the device mnop by means of which the thread of the weight can be adjusted to the centre by the screws n and o.

The centrum motus is also the centrum divisionis. The divisions are said to have been made according to the calculated chords.

Fig. 3 is one of the brick piers for the transit instrument. AB is an iron bar passing through the wall into which the devices for the bearings are screwed. The front of mn is shown in Fig. 4. The brass plates are screwed into it in the four corners xyzw. AB = 24 inches; BC = 12.5; BD = 6 feet 6 inches; DE = 3 feet.

The mural quadrant in Bird's description, the model in the British Museum and the quadrant at Greenwich[200] are so constructed that only half of the weight of the

200 In 1749–50, Bird had constructed an all-brass 8-foot mural quadrant for Greenwich Observatory. According to Bird's entry in the Oxford National Biographical Dictionary, a half-size wooden model of this was prepared and lodged in the British Museum, but is now lost.

movable telescope has been eliminated. The other half presses down upon the centre. Mr Hornsby has completely removed this pressure so that his telescope moves without the slightest friction by means of the two levers shown in the figure.

The 9th

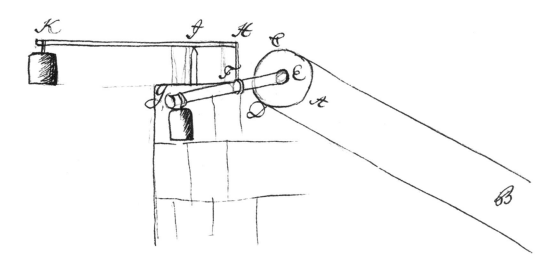

AB is the plate to which the telescope is attached. CD is the central plate around which the movement takes place. To this plate a round iron bar efg has been fastened. G is the weight, and F is the centre of suspension which is carried by another lever HIK with the weight K. EF = 3 inches; FG = 6"; HI = 6½"; and IK = 12½".

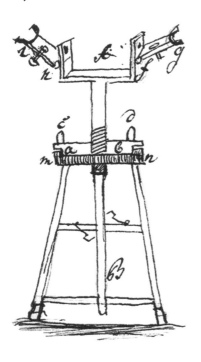

By day as well as by night Mr Hornsby allowed me to make several observations with the transit instrument. And I was thereby convinced that it is not impossible to make observations of an accuracy of about 0.2" with such a high magnification, which I used to consider as boastfulness rather than reality.

To complete my description of the apparatus for reversing the transit instrument, I must add the following:

1°) In the disk mn another disk ab can be turned around by the two handles cd. Thereby the bar AB can be screwed up without revolving, which would make the mounting of the transit instrument impossible.

2°) On the side plates of A the hinges fg and hi can be moved at h and f. They can be moved up to and away from the axis. Moreover they can be fastened by means of two screws.

I should like to add the following practical remarks:

a) All the perpendiculars and filaments were made of silver thread.

b) All the divisions are – if possible – read off by the same light, viz. by a wax taper; the microscope is covered by a piece of thin oil paper.

c) The sunrays are kept away from the instruments. The axis of the transit instrument in front of the cube is covered with a broad piece of paper, and this is better than Maskelyne's method: to cover the entire axis with paper. In front of the mural quadrant, towards the south, a thin mahogany plank is placed (1 line thick). *The rays of the sun must also be kept away from the eyepiece and the nonius.*

d) In order not to make the instrument too heavy the [Lagselen? = ?] for the weight and other things are made of mahogany.

In the room of the mural quadrant there were 4 thermometers with a Fahrenheit's graduation: two by Bird and two by another artist. And to my great surprise I saw that they agreed without any notable difference.

A separate small, round building had been

The 10th

erected to house an equatorial instrument *by Bird* [201]

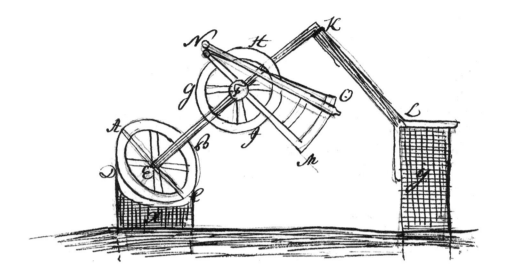

x and y are two brick piers: x has approximately the same altitude as the equator, according to which the brass circle (of about 2 feet diameter) can be adjusted; perpendicular to this a solid iron bar of 2 inches square has been attached. It is carried by the iron bar KL which is supported by the wall y. EK can be adjusted by screws on KL. GHI is a circle of about 2 feet diameter, around its centre F can be rotated a sector NOM of 3 ½ feet radius fitted with a telescope and a nonius.

201 This instrument is not listed in Howse 1994, but is discussed in Guest 1991, p. 245. He mentions that Bird had died before this equatorial sector was finished (which means that it cannot have been erected long before Bugge saw it), and that it "never was good for much" and by 1830 was considered "nearly useless".

Between the lowest part of the house M and the movable roof there was a row of brass rollers a,b,c,d,e etc. moving in two brass grooves. By means of this arrangement it was possible to move the entire roof N with one hand.

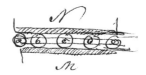

Snow and rain had fallen into Mr Hornsby's large observatory. He had then provided all openings with wood strips a and b, and found that this had the desired effect.

Mr Hornsby lectures on Experimental Physics in the museum where he has a beautiful collection of physical instruments of his own.[202] Among other experiments the following deserves special attention.

Prospect from the North

202 Since the late 17th century, experimental philosophy was taught in the (Old) Ashmolean building along Broad Street, which now houses the Museum of the History of Science. Hornsby lectured here from 1763 onward. See Turner 1986 and Bennett, Johnston, Simcock 2000, pp. 18–20.

Oxford
October 1777
The 11th

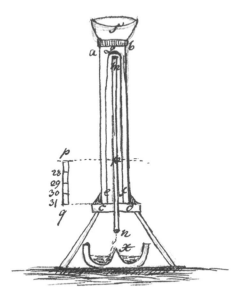

abcd is a glass tube with a diameter of 1 to 2 inches and a height of 33 to 34 inches. In its centre is another tube mn, open at its outer ends, 30 to 32 inches long and 1 to 2 lines in diameter. Around this is placed a detachable tube egf of 4 to 5 lines in diameter and so long that it fits exactly to the top of m.

From this point the ordinary barometer inches are marked. This is marked on the tube, but for better illustration I have shown it in the figure between p and q. When mercury is poured into the funnel s, it fills the tubes egf and mn. But some of the mercury will by itself run into the outer tube until mp or qp are exactly equal to the barometric height. The cone x is so placed to prevent the outflowing mercury from splashing.

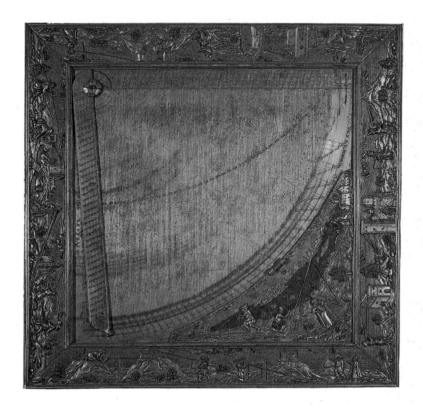

Geometrical quadrant by Christoph Schissler, Augsburg, 1579. 346 x 346 mm. Museum of the History of Science, Oxford. • The border is decorated in relief with scenes of surveyors in action. ▶

In the lecture room the benches were placed amphitheatrically in a very good way, thus allowing the entire audience to follow the experiments. In Fig. 1 the benches are seen in profile.

Fig. 2 is the ground plan of the lecture room. M is the table where the experiments are carried out. ABC is the first, DEF the second, and GHI the third row of benches.[203] He gave me a printed draft of his lectures. The astronomical lectures are divided in 14 periods, each period takes seven quarters of an hour. The experimental physics is divided in 26 such periods. During a period many unusual experiments are carried out.

I saw the theatre which is a beautiful and impressive building constructed by Ch. Wren.[204] It is used for the same purposes as the largest lecture room in other academies.

I saw the Bodleian Library[205] where there is a gold quadrant with a radius of 1½ Kvarter.[206] Above the library is a very large gallery with portraits of famous men.

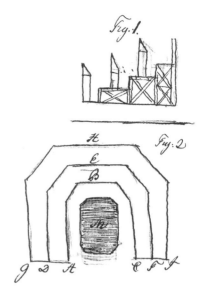

203 It is not certain whether this amphitheatre was in the basement or on the ground floor; for the arguments, see Bennett, Johnston, Simcock 2000, p. 20.

204 The Sheldonian Theatre, next to the Ashmolean building, was erected in 1664–8 to a design by Christopher Wren from funds donated by Gilbert Sheldon (1598–1677), Chancellor of the University of Oxford.

205 The Bodleian library was founded by Sir Thomas Bodley in 1602. Apart from books and manuscripts, it housed other types of objects including those mentioned by Bugge: a quadrant (see next note), portraits and original statuary, some from the Arundel collection, which had arrived in 1667, some given by the Countess of Pomfret in 1755.

206 Gilt copper geometrical quadrant, made in 1579 by Schissler of Augsburg, reputedly presented by Sir Thomas Bodley himself. It was one of the sights of the Bodleian Library, where visitors were told that it was made of gold. The Kvarter equalled 6 inches; the radius of the engraved quadrant is indeed some 9 inches.

October 1777
Oxford
The 11th

Of Danish men I found Griffenfeld,[207] and Tycho Brahe, and of English Mathematicians: *Wallis, Flamsteed, Halley* and *Bradley.*

Below, in two separate rooms, were kept the *Arundel Marbles, the Pomfret Statues, Busts and Marbles.*

In the lowest floor there are Schools. They are smaller lecture rooms for different purposes. Among the colleges especially the following deserve to be seen: All Souls College, Christ Church, Magdalene College, Queens, and University College. A more detailed description is found in: A new Pocket Companion for Oxford, or Guide through the University.

I saw the Museum Ashmoleanum. On the lowest floor the professors give lectures in experimental physics, chemistry, and anatomy. On the second floor the real museum is situated, besides the Ashmole, Lester, and Wood libraries. As for the Natural History it does not contain much of interest. There are costumes and other curiosities from the South as well as from Otaheite. A man dressed for war, and another figure in mourning.

Mourning dress from Tahiti. Pitt Rivers Museum, Oxford. One of the ethnographic specimens brought back from Captain Cook's voyages that Bugge saw in Oxford. ▶

207 The Danish statesman Peder Schumacher (1635-1699), after 1673 Count
     Peder Griffenfeld, was at Oxford, Queen's College, from 1657 to 1660.

They correspond exactly to the drawings found in Cook's Voyages.[208]

Probably the biggest natural magnet in Europe is found in this museum;[209] the view from above is as follows.

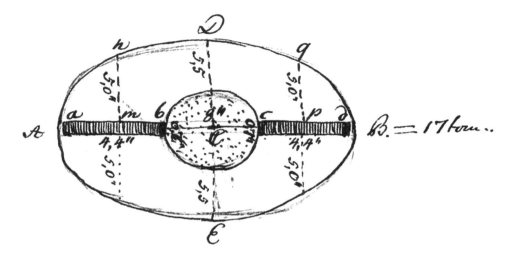

The greater diameter AB = 17 inches; the smaller one DE = 12 inches. The fittings ab and cd = 4.4". The diameter of the circle bc = 8". This circle has been cut out, and there the magnet itself is visible, whereas the rest of the ellipse is covered with a brass plate ADBE. The fitting is 0.7" thick. The ordinates mn and pq = 5.5". This magnet is a donation to the University from the Countess of Westmorland; and it carries 150 lbs.

208 Johann Reinhold Forster and his son George joined Captain Cook's second voyage of discovery to the South Seas in 1772–75 as naturalists. In 1776 they presented to the Museum about 150 ethnographic specimens, including a spectacular Tahitian mourning dress, now in the Pitt Rivers Museum, Oxford. (Otaheite is the historic name for Tahiti). MacGregor 2001, pp. 32–35, with colour photo of the dress.

209 This imposing natural magnet, displayed in a mahogany case and mount, elegantly highlighted with a gilt ducal coronet surrounding the magnet, was presented to the Ashmolean Museum by the Countess of Westmorland in 1756, and is now in the Museum of the History of Science. An engraving published to mark the presentation is reproduced on page 127.

In profile this magnet has the following appearance.

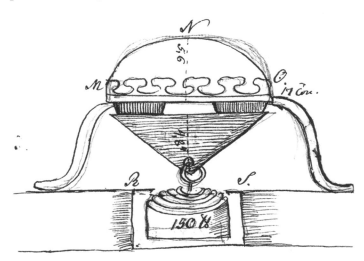

The drawing gives a very satisfactory impression of the whole arrangement. I only want to point out that the upper part MNO was covered with red leather and that the rest of the magnet was decorated like a crown. The weight was hidden in a hollow space, and at the very top it was made up of lead rings with a decreasing diameter.

With the money left by Doctor Radcliffe the following buildings have been erected: 1.) The Radcliffe Library. 2) The Infirmary where the patients are provided for and treated by subscription. 3.) The Observatory with instruments to a total amount of £ 20,000 or 100,000 rigsdaler in Danish currency. It is anticipated to be enough for constructing a building for a collection of physical instruments as well as a reading room.

The 12th

[From here to fol. 70 verso, Bugge gives detailed data of astronomical observations. For a transcript, see appendix 4].

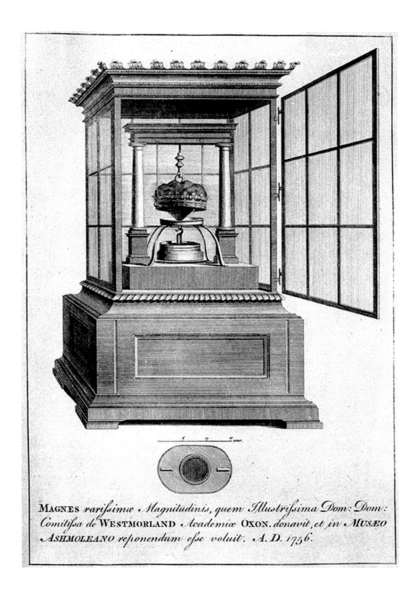

◀ Engraving published to mark the presentation of the large magnet to the Ashmolean Museum in 1756. Museum of the History of Science, Oxford.

October 1777
Oxford
The 13th October

*NB. Prof. Sir John Pringle has
told that Ramsden, Dollond
and Nairne knew how to
grind parabolic mirrors. Short
was probably the first and
there is an account of it in
the last volume of Philosoph.
Transacts.* [212]

The very large 12-foot reflecting telescope of which de la Lande reports[210] that it is said to have been in the possession of the Duke of Marlborough, has been donated to the Oxford Observatory.[211] However, it has not yet been unpacked, so it was impossible to have a look at it. A similar one is said to have been constructed by Short and sent to Spain.[212]

Not without regret did I leave the Oxford Observatory which is no doubt the best in Europe, both as regards the arrangement and the instruments. Professor Hornsby most courteously assured me of his friendship and his correspondence.

Nearly all the colleges at Oxford are in the Gothic style; they must have cost enormous amounts of money.

A new bridge is under construction towards London. It is too narrow and in a bad taste. It consists of various arcs,

210 The French astronomer Joseph-Jérôme le Français de Lalande (1732–1807) visited England in 1763. In Lalande 1771, par. 2422, he discusses James Short's telescopes, including the largest version of 144 inch (12 foot) focal distance, 'of which he only made one, which in 1763 was in London at Marlborough's residence, but it was dismantled, and nobody made use of it.' This information is not in Lalande's travel diary, published in French (Lalande 1980) and English (Watkins 2002).

211 George Spencer, fourth duke of Marlborough (1739–1817), had a private observatory, first at St. James's, Westminster, then at Blenheim Palace near Oxford; Howse 1986, pp. 66 and 76. The donated reflector, made by Short in 1742, is in the Museum of the History of Science, Oxford, inv.no. 98721. Bugge's note shows that it did not come to the observatory as late as 1812, as had been thought previously; for details, see Guest 1991, p. 242.

212 According to Short's entry in the Oxford Dictionary of National Biography, among his commissioned instruments was an 18-inch aperture Gregorian reflector for the King of Spain, completed in 1752 and costing £1200.

213 The surgeon and physician John Mudge (1721–1793) won the Royal Society's Copley Medal for his treatise on making reflecting telescopes, printed as Mudge 1777.

half-circles and ellipses.[214]

The town is situated in a very pleasant place, surrounded by well-cultivated land within a radius of ½ to ¾ miles (Danish). On the opposite side of Oxford a new highway with 4 bridges has been constructed across the meadows. The first mile is said to have cost £ 5000.

After my return to London I attended a meeting in the Royal Society for the Encouragement of Arts, Manufactures and Commerce in their new building on Adelphi.[215] It was half past seven p.m. before 7 members were assembled, which is the quorum. *the entire meeting did not have more than 14 persons*. The hall is arranged as shown.

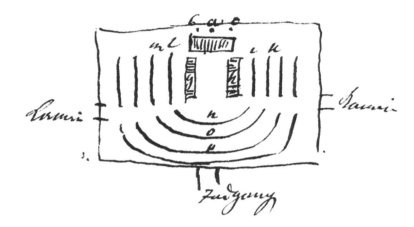

214 Swinford Bridge over the river Thames was completed in 1777 as part of a new turnpike road, with classical arches, cut from a rather soft stone. Text and photo in Atterbury 1998, p. 27.

215 This time, Bugge misnamed it English Agricultural Society, having called it London Agricultural Society when he visited the Society's Repository on 24 September.

October 1777
London
The 15th

a. is the seat of the president or vice-president. b. and c are the seats of the secretaries. g and h are two tables where the other vice presidents are seated. ml, nik, o, p are benches for the members.

This evening extracts from letters from different committees were read. A letter from a man applying for the post of treasurer. After about half an hour the Society had nothing else to do. Among other things there were some reports or notifications from several other societies in the counties.

I did not like the way matters were dealt with. There were no debates, and the vice-president, Mr Hooper,[215] does not seem to be a very fluent speaker, judging from his few addresses to the Society.

At Mr Adams [216] I saw a new transit instrument which he is making for the King's Observatory at Richmond.[217] It is of the same design as the transit instrument at Oxford.

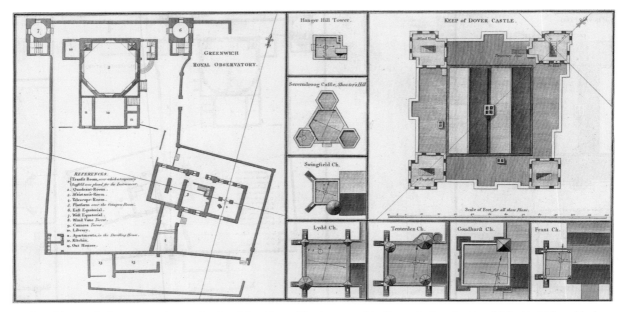

▲ *Plan of the Royal Observatory, Greenwich. Plate XII in General Roy's report of the Triangulation of 1788, published in Philosophical Transactions of the Royal Society of London vol. LXXX (1788).*

When I was in Greenwich on October 18th, I made the following drawing of the observatory.[219]

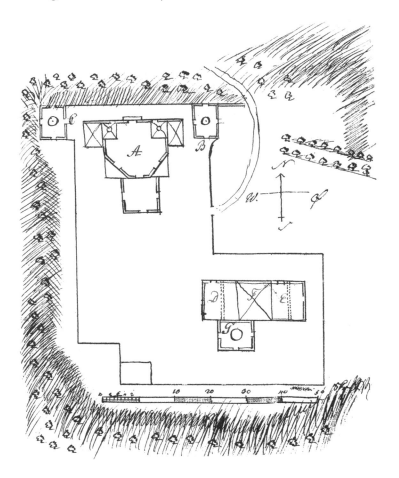

216 No particulars could be found on this man except his first name, Edward, see Trueman Wood 1913, p. 77.

217 George Adams junior's shop was at 60 Fleet Street. Bugge does not mention it, but one assumes that he now bought the set of drawing instruments by this maker, listed earlier (see appendix 3).

218 As discussed earlier (note 134), it is not clear whether it was ever delivered to Richmond.

219 Bugge gives an incorrect orientation of the observatory building in the south-eastern corner, which should be turned some 20 degrees to the west, as can be seen from the engraving reproduced opposite. Bugge's explanations are on fol. 87 recto and following.

The front of the Greenwich Observatory, seen from the Hospital.

I visited Mr Russel,[220] who showed me several of his instruments. A spirit level, sent by one of his friends. It is very suitable for smaller operations.

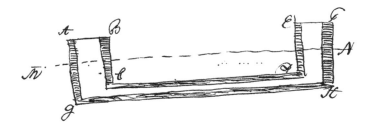

220 William Russell, whom Bugge had visited earlier on 5th October.

It consisted of a small tube of mahogany, 6 inches long, communicating BC and ED, more than 1 inch long. Into the upright tube BC fits a mahogany stopper P with a quadrangular brass plate R with a hole in it. Into the other upright tube ED fits the stopper Q with the brass piece S, in which a circle has been cut and fitted with a horizontally distended hair.

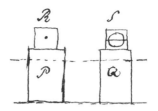

The tube is filled with mercury to the horizontal level MN whereupon P and Q are floating. In this way the horizontal level can be determined.

Mr Russel's equatorial instrument has no horizontal plate, but only an equatorial or rectascentional plate and a declination plate, each with a radius of 2 ½ inches. It supports a 10 inch reflecting telescope with an aperture of about 3 inches. It is fitted with an excellent objective micrometer by means of which Russel had determined the diameters of Jupiter and the Moon with an accuracy of 1 second. He

FACIES SPECULÆ SEPTEN.

▲ *The north face of the Royal Observatory. Etching by Francis Place, c. 1676. This print, published shortly after the founding of the Observatory, shows the view from the river Thames, with the two equatorial turrets on the sides.*

October 1777
London

showed me his observations. The instrument was made by Bird.

Furthermore, he showed me a small, handy 3 foot telescope with matching support. AB is the telescope, which is fastened by a brass band CD. This can be opened by means of hinges, and clamped with a screw (see Fig 1). By means of the plate at E it can be adjusted to any desired height and then be fastened by a screw.

The cylinder F fits into the wooden spigot on the support where it can be fastened by the screw G.

The micrometer consists of two hairs at right angles, and two other hairs intersecting each other at an angle of 45°. In this way it is possible to

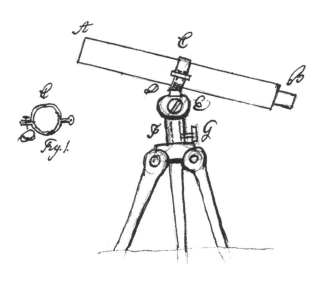

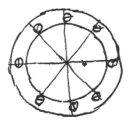

observe the Differentia rectascensionis et Declinationis.

He also had an excellent spirit level, 6 inches long, made by Bird, and constructed in the same way as the Centre or Trial Telescope described by Smith in his Opticks. He places the spirit level on the quadrangular plates as shown in the following figure.

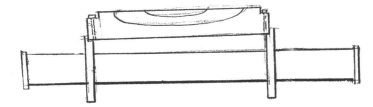

Moreover I noticed that the spirit level could be adjusted by means of two screws placed at each end of it.

An *artificial horizon* consisted of a glass mirror plate of 4 inches square. Its upper and lower plates were exactly parallel. The glass plate rested on a brass square ab, to which it was clamped into a frame so that the glass was elevated at the bevelled facets.

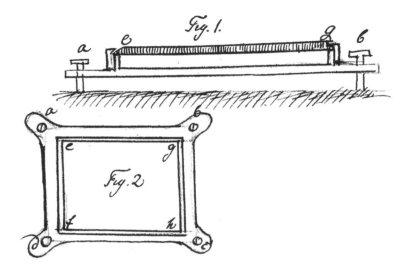

Fig. 1 is the profile and Fig. 2 is the plane. abcd are 4 adjustable screws. efgh is the brass frame, and eg in Fig. 1 shows the mirror glass or the artificial horizon.

At Mr Russel's I saw the following curious books:

The Antilogarithmic Canon being a Table of number of Eleven Figures; by *James Dodson*. London 1742.

Nb. Benjamin [220]

The Calculator being correct and necessary Tables for Computation by Dodson. London 1747. Very useful tables for computation of interest, measure & weight, computation of circles and spheres, sine tables [221]

---

221 We cannot explain this marginal note. It is not the first name of the author of both volumes (Dodson 1742 and 1747), the mathematician and actuary James Dodson (c. 1705–1757).

which de la Caille[222] seems to have used.

Robert Hook's Works, containing many good ideas, in particular concerning instrument making.[223] These works are very rare.

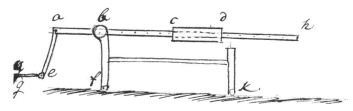

Finally, Mr Russel showed me a little balance by Ramsden for weighing coins.[224] The basic idea is that the coin, one guinea, ½ guinea, ½ crown, etc., is placed on the plate ge, and the poise cd is moved along the bar bh to the mark corresponding to the coin under consideration, one guinea, ½ guinea, etc. If the coin is in order, the balance will be in equilibrium. The unit of division is 1 shilling. In Fig. 2 bf and the suspension is seen from the front. In Fig. 3 ae and its suspension is also seen from the front.

e., f. and k. are joints so that

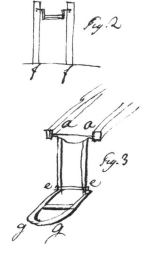

◀ *Coin balance, with label on box signed Ramsden Haymarket London, c. 1770-1773. Box width c. 75 mm. Whipple Museum, Cambridge.*

222 The astronomer Abbé Nicolas Louis de Lacaille (1713–1762). who in 1750–1754 studied the southern hemisphere from the Cape of Good Hope, reputedly observing more than 10,000 stars.

223 The natural philosopher Robert Hooke (1635–1703), whose Posthumous Works appeared in 1705, edited by Richard Waller.

224 The folding guinea balance seen by Bugge is a version of the common Lancashire folder; Ramsden took the basic idea and improved on it. A similar Ramsden balance is illustrated above and discussed in McConnel 2007, pp. 184–185. See also Camilleri 2001, which includes one by Ramsden, and Crawforth 1979, p. 57.

October 1777
Cambridge

The 23rd I lodged at The Rose.

the instrument can be folded into a wooden box 3 inches long and 1 inch thick.

On October 23rd I went to Cambridge together with Mr Olivarius, M.A., from Copenhagen. Two Englishmen, Mr Waterson and Mr Yale, both Fellows of St. John's College, were kind enough to show us the King's Chapel, the most beautiful gothic church in Europe, the Senate House (or the great lecture room), the University Library, and all the Schools Colleges and Halls. A tolerably detailed description is found in *Cantabrigia depicta* , or A Description of the University and Town of Cambridge. However, it is not as complete as the above-mentioned description of Oxford.

At Trinity College a small astronomical observatory has been fitted up in a sort of tower,[225] but for the moment it did not contain any astronomical instruments.

At St. John's College is the observatory

The 'Danish Geographical Instrument'. Engraving from Thomas Bugge 1779a. This instrument was made by Johannes Ahl c. 1762 and used by Bugge for the triangulation of Denmark. ▶

225 Howse 1986, p. 67, lists this as in use c. 1707-1797, and mentions two main instruments: a sextant by Rowley, London and a clock by Street, London.

where Dr. Ludlam[226] has made his observations.[227] The instruments belong to St. John's College and comprise the following:

a.) a transit instrument with a 3 foot telescope with a device for observing at right angles to the axis.

b.) a movable quadrant of 22 inches radius. It is placed on a stone. On the whole the arrangement is the same as that of the Danish geographical instrument.[228] Its plates are very thin and have no counter or cross plates so that the instrument is almost flexible because of its own weight.

c.) a good regulator or astronomical clock with a gridiron pendulum.

d.) a 3 foot reflecting telescope with a support which can be placed in a window or on a wooden support made especially for this purpose. The telescope was fitted with an objective micrometer.

The observatory has been erected on a small tower.

226 William Ludlam (1717–1788), mathematician and writer on theology, became a fellow of St John's College in 1744.

227 The observatory atop the Shrewsbury Tower of St. John's College had been planned and funded by Richard Dunthorne, who had also donated the instruments. He observed the transits of Venus in 1761 and 1769, and prepared new lunar tables. William Ludlam published astronomical observations made there in 1767–8. Howse 1986, p. 67, lists it as in use 1765-1859, and mentions two of the four instruments mentioned by Bugge: a transit instrument by Sisson, London, now in the Whipple Museum, Cambridge, and a regulator by Shelton, London, now in St John's College Library, see illustration. The quadrant that Bugge saw has probably not survived. It is not the Bird quadrant from St. John's Observatory in the Whipple Museum, inv. Wh. 1084, illustrated in Bennett 1987, plate 96.

228 The 'Geographical Instrument' or 'Eckstrøm Circle', made ca. 1762 by Johannes Ahl, used by Bugge for the triangulation of Denmark, is now in the Kroppedal Museum in Høje Taastrup. See the engraving on the opposite page; see also Nielsen 1989, pp. 30–32, and Thykier 1990, pp. 191–193 and 580, n. 4.

▲ *Regulator in veneered mahogany case, made by John Shelton, London, c. 1764. Height 190 cm. It was used in St John's Observatory. St John's College Library, Cambridge.*

October 1777
Cambridge

The 24th

I was told that all pillars and walls supporting an instrument were erected upon vaults.

I looked for Doctor Shepherd,[229] Plumian Professor of Astronomy, at Christ College where he lives and holds a tutorship, but at that time he was in London *Sherwood[230] St. N°. 1 Golden Square, London*. However, his servant showed me into the observatory,[231] a small house, standing by itself between the house and the garden.

The instruments were not placed on the ground floor nor on the first floor, but on the second floor about 20 to 24 feet above the ground. Dr. Shepherd has had it erected at his own expense. It contains

a.) a 3-foot transit instrument with an axis of 1 foot.

b.) a movable quadrant with a radius of 20 inches with a double graduation. A counterweight for the telescope. It can also be placed in a horizontal

---

229 Anthony Shepherd (1721?–1796) was elected Plumian professor of astronomy at the University of Cambridge in 1760 and FRS in 1763.

230 Bugge wrote Sherrot.

231 Howse 1986, p. 67, lists this as in use c. 1760–1780, and mentions four main instruments: a transit instrument, a mural quadrant, and clocks by George Graham and Robert Allam, both of London.

position, and to its back is attached a spirit level with a length equal to the radius of the quadrant. By reversal the spirit level (ab) can be placed parallel to the upper plate by means of screws.

Fig. 1 is the above mentioned horizontal position, and Fig. 2 shows the vertical position of the quadrant. It is then hanging in the same way as ab.

c.) a regulator with a gridiron pendulum, made by Graham.

d.) a Dollond telescope with 3 object glasses, 3 ½ feet long and with an aperture of 4 inches. The support was very stable and of the construction normally used for reflecting telescopes.

In his room I was shown:

1.) an astronomical clock with a pendulum made as a tube, and I have read somewhere that these clocks go quite well, but I do not remember where I have seen it.

2.) a telescope with the device for motus parallacticus which is described by Bernoulli in his Lettres Astronomiques, pag. 118-119.[232]

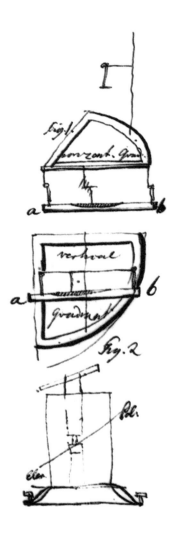

---

232 Bernoulli 1771.

October 1777
Cambridge

For this telescope Dr. Shepherd had an objective micrometer. 3.) microscope. 4.) a concave mirror. 5.) a convex mirror. 6.) a hygrometer. 7.) a barometer and a thermometer by Bird. None of these instruments were especially remarkable.

A very nice tool for proving the elliptical movement of the sun.[233] It is constructed as shown in the figure.

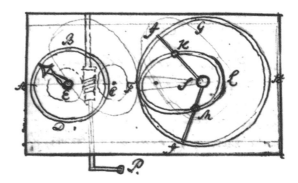

ABCD is a circle divided into 24 or 365 parts. FGHI is a circle representing Zodiacus or Ecliptica and it has been divided into the 12 signs of the zodiac. The sun is placed in its centre, and this is at the same time the focus of the earth's elliptical

Cometarium as designed by
Desaguliers, made by Stephen
Demainbray, c. 1755. 50 x 39 x 11.5 cm.
Science Museum, London.
It represents the path of a comet
around the Sun, the small dial
showing the passage of time. ▶

233　What Bugge evidently considered a machine showing the elliptical motion of the Earth about the Sun, must in fact have been a cometarium, constructed along the same lines as the Desaguliers and Ferguson cometaria; for an example see the cometarium illustrated above, discussed in Morton and Wess 1993, pp. 160–1. With thanks to Martin Beech, University of Regina, Canada, for his comments on this description and drawings.

orbit KLMF. Through S has been stuck an iron pin SK with a small, movable globe representing the earth, falling down into the hollow, elliptical orbit FKLM. When the handle P is turned the earth will remain in its elliptical orbit, and the other end of the pin I follows the Zodiac. On the ellipse the areæ temporibus proportionales have been marked.

The interior construction is as follows:

October 1777
Cambridge

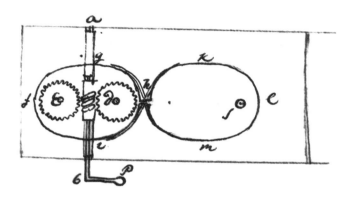

ab is an endless screw with a handle P. This screw sets in motion two wheels with 24 teeth. E moves the pointer, and D moves the focus of the ellipse fghi.

October 1777
Cambridge

Around it is a rope which sets the other ellipse hklm in motion, and one of its focuses S moves the sun.

During terms or regular reading times at the University many subjects within theology, jurisprudence, and philosophy are discussed. These discussions take place every afternoon from 2 to 3½ in the various schools or lecture rooms, under surveillance of moderators. Normally one college is debating against another.

The course of studies in mathematics is as follows: 1st year. algebra, Euclid's 1st and 5th book; 2nd year. The rest of Euclid's first 6 books, and sectiones conicae, astronomia spherica; 3rd year. mechanics, hydrostatics, pneumatics; 4th year. astronomia physica, based on Newton's Principia.

It takes a very long time to obtain the various university degrees. Baccalaureate, 2 years. Master of Arts, 5 years. Bachelor of Divinity, 9 years. Doctor of Divinity, 17 years. Doctor of Law, about the same time. Doctor of Medicine, somewhat shorter. To become a Doctor of Law costs 40 pounds.

It is very curious that the professors give almost no lectures. They have nothing to do but to examine. Thus Dr. Shepherd spends most of his time in London, and professor Waring[234] spends most of his time in the country. But besides there are lecturers or readers in all sciences at each of the colleges.

After my return to London Mr Russel introduced me into the Astronomical Club where I met: Mr Aubert; Dalrymple; Cavendish; Dr. Aberdeen; Nairne; and Lord Mahon; Dr. Shepherd, and others.[235]

London

I visited Mr Lever's Museum, called Leicester House, situated at Leicester Fields.[236] His collection is probably the largest in Europe as regards birds and animals. The museum consists of 8 rooms, very nicely arranged and decorated. Mr Lever told me that Captain Cook is now collecting objects for him on his third voyage, and so do many Englishmen in the East and West Indies. Thus he hopes

234 Edward Waring (c.1735–1798) was appointed Lucasian professor of mathematics at the University of Cambridge in 1760.

235 We found no information on this Astronomical Club, which was presumably informal, and whose members evidently included the merchant and astronomer Alexander Aubert (1730–1805), whose observatory Bugge was later to visit; the hydrographer Alexander Dalrymple (1737–1808); the natural philosopher Henry Cavendish (1731–1810); probably the agriculturist and political economist James Anderson (1739–1808), alias Aberdeen; the Cambridge professor Anthony Shepherd; and Nairne and Lord Mahon, whom Bugge had met earlier.

236 The naturalist Sir Ashton Lever (1729–1788) opened his private museum on Leicester Square in 1774. Also known as the 'Holophysikon', it was London's most extensive museum of natural history, ethnography and miscellany. After Lever's death it re-opened elsewhere in London as 'Museum Leverianum', before the collection was sold by auction in 1806. See Altick 1978, pp. 28–32, and King 1996.

to double his collection within 3 years. The museum is open every day from 10 a.m. to 4 p.m., and the entrance fee is ¼ guinea. A one-year subscription costs 2 guineas.

I went again to see the watchmaker Mr Cumming, New Bond Street, where I carefully examined his barometric clock.

It is worth noticing that the dial has an hour hand A *and a minute hand B* indicating sidereal time, and a number of tiny, gilt points, dotted on the glass plate, which indicate the culminations of the stars. Another hand, together with the second hand, indicates mean time.

The way in which the pencil P delineates the barometric height
on the white parchment plate, working for three years and making one revolution during the course of a year, is the following:

AEGL and DFHM are two barometers communicating with
the same reservoir IKLM. The diameter AB = CD is 2 inches. E
and F are ½ inch. The diameter LM is about 4 inches. On top of
the mercury IKLM an ivory plate IK is floating. The stick NP, attached to the plate, moves the pencil. At R and S 4 friction rollers
are placed. As the mercury falls and rises, the pencil P goes up
and down,

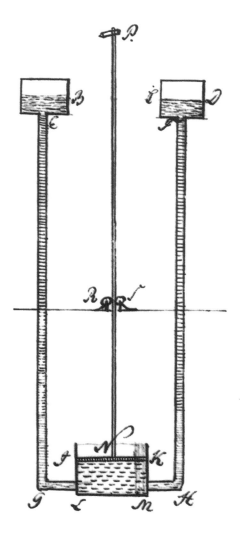

October 1777
London

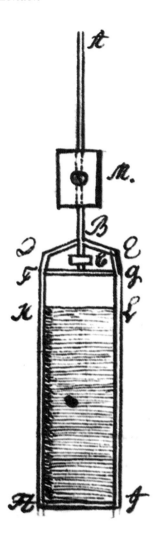

thus describing on the parchment plate the curved line representing the barometric height. For such a barometric clock Cumming charges 500 pounds sterling.

At Mr Cumming's I saw a very curious thing; namely the only existing clock with a mercurial pendulum, invented and constructed by Graham. Cumming told me that he bought it at an auction after Lord Macclesfield.[237]

ABC is the pendulum rod made of brass. It is about half an inch wide and 3 lines thick. By means of a brass frame BDHIE it carries a glass cylinder FHIG with a brass cover FG, with a disc C. KL is the surface level of the mercury.

The expansion of the brass rod is counterbalanced by the

237 Bugge does not seem to recall that three weeks earlier he had already made a drawing of the mercury pendulum on the Graham regulator which had been in the Earl of Macclesfield's possession (62 recto). With 'after', Bugge's presumably meant 'from'. We do not know which auction he refers to.

expansion of the mercury. The adjustment can be done by changing the quantity of mercury or by moving the weight M up or
down, by which the centrum oscillationis is changed.

On October 31st Dr. Maskelyne came to see me, and he took
me to Greenwich. But as he asked me to dine with him on November 5th and to have a better look at the observatory, this will
be described in detail later on.

The 2nd November I was asked to dine with Mr Alexander
Aubert Esq. in his summer house Cowbridge near where he has a
very nice observatory.[238]

First of all he has provided very solid foundations for the instruments. On a layer of gravel and a layer of fine sand he dug
down 6 feet and laid a foundation abcd of bricks and liquid lime.
The surface is concave in order to prevent the slab of stone dc
from breaking. Upon this stone he erected two other stones.

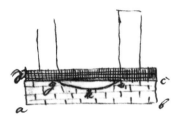

◀ Portrait of the merchant and
astronomer Alexander Aubert
(1730-1805). Stipple engraving by
J. Chapman after S. Drummond,
reproduced in European Magazine
(1798). Bugge visited Aubert's private
observatory at Deptford, and his
description with drawings is the most
detailed information we have of this
observatory

238 Bugge wrote Cowbridge in error: that was where Pigott's observatory
was, as noted on fol. 50 recto. Aubert's observatory was on Loampit Hill,
in Deptford near Greenwich. On Aubert, see the entry in the Oxford
Dictionary of National Biography, and Aubert 1951–52. His observatory
is listed in Howse 1986, p. 69, as in operation from before 1769 to 1788,
when he moved to Highbury near Islington, London; Howse 1986, p. 74.
Howse lists five instruments, all before 1769 and by London makers: a
transit instrument by Bird with optics by Dollond, clocks by Shelton (2) and
Ellicott, and a mural quadrant by Bird; see also Howse 1994, pp. 214–215.

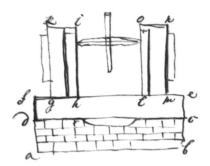

He left it like this for 3 to 4 months to allow the foundations to settle. dcef is the horizontal foundation stone. ghik and lmno are the two pillars made from one stone cut in two equal pieces so that each has the same expansion and contraction.

The transit instrument is placed in the middle. At ih and ol are two clocks. The mural quadrant is placed on the eastern side xg, and it can be moved to the western side nm.

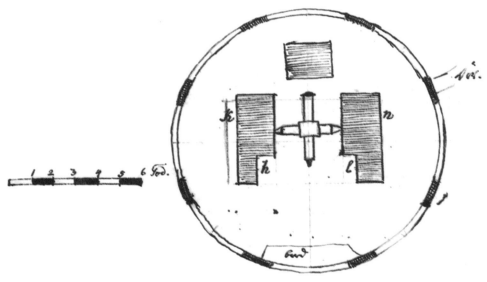

The plan is shown in the opposite figure, where *h* is the clock made by Shelton. It is hanging directly on the wall, and the case is hanging separately to protect the clock from dust. l is a counter. k and n are the sides of the pillars on which the mural quadrant has been erected. I will now give a description of the instruments.

1°.) *The transit instrument* is fitted with a 3 ½ foot achromatic telescope with 3 glasses. The aperture is 3 ½ inches. The axis is 2 ½ feet. The bearings are made of yellow or black bell metal. The pivots are perfectly cylindrical and made of *steely white bell metal*. It is covered by a piece of lead, lined with cloth and coated with tallow and oil. It is very important that the axes of both pivots are aligned and completely cylindrical. The friction is eliminated in the same way as described at the Oxford Observatory. The method of lighting is also the same. It is an advantage that the lamp is large, and

November 1777
London

that the smoke flue is turned backwards. The illumination plate is an elliptical ring. Its minor axis has the same diameter as the aperture of the glass, and it is black on one side and white on the other. Mr Aubert had two filaments. He had two excellent meridian marks, one towards the south, the other towards the north. The first one, placed low in front of houses and trees, was an enamelled white 3 inch disc on a black ground. The other mark was placed in the open and consisted of a hole in a black plate.

As Mr Aubert knows the exact distance between the collimation line of the transit instrument and the collimation line of the mural quadrant, he has inserted an arm ab, on which the distance has been drawn as a white line. This is the meridian mark of the quadrant.

2°.) *The mural quadrant* has a radius of 4 feet, but it is graduated in the same way as the 8- foot mural quadrant at Greenwich. It is the most perfect and accurate instrument of its size.

According to Bradley's and Bird's construction half the weight of the movable telescope has been eliminated. Mr Aubert has found out how to eliminate the remaining friction by means of the device mentioned in my description of the Oxford Observatory. Mr Aubert finds the relevant weights in the following way:

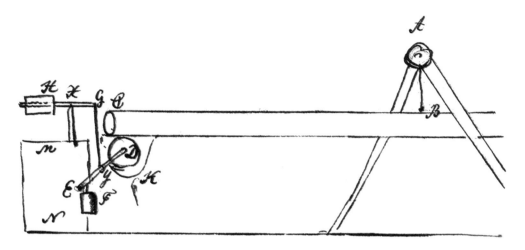

On a table a tripod A is suspended.[239] From this, one end of the telescope is suspended in Bird's centre of gravity. On MN rests the lever HG, connected to which is the lever ED. The balance arm ED is correct in relation to the hypomochlion[240] and to the weight F when the plate DK is vertical. Errors can be corrected by increasing the weight HF [?] or by changing the point y.[241]

239 Probably Bugge means 'placed'.

240 Fulcrum, point of support.

241 The last line is unclear in the manuscript and may be incorrectly translated.

London
November 1777

When the telescope CB is horizontal, the lever HG is correct. In other cases the weight H has to be increased or diminished. It could also be corrected by moving the weight. The graduation and the micrometer screw are so perfect that they show exactly the same parts of the circle arc. But then care must be taken always to move the screw to the same side, because of the clearance between the screw threads and the screw marks which would cause an error.

Furthermore, Mr Aubert has made another improvement on Bradley's or Bird's levers in order to eliminate the weight of the telescope.

It is obvious that one lever, PQ, is not able to eliminate the weight of the movable telescope in all positions. It supports the telescope in the horizontal position AC, but when the telescope is in position CB, it does not support nearly as much,

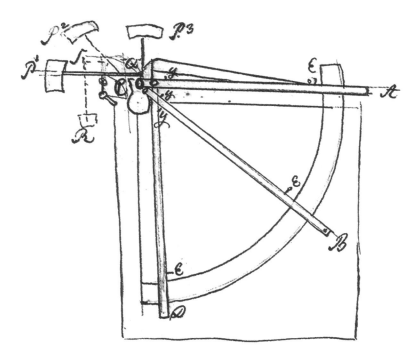

and finally in the vertical position CD of the telescope nothing of its weight is eliminated, because the weight P follows the telescope and has the positions P^1, P^2, and P^3. Therefore, Bradley has placed the second lever, RS, in a contrary position, so that it does not support the telescope in its horizontal position, but from this position its supporting effect starts and increases until the telescope is vertical in CD, where the lever RS becomes horizontal and supports the telescope. Bradley and Bird had placed it in such a position that it began to support from E or from the same

November 1777
London

point as the lever FQ. But then it is obvious that when the telescope is vertical, the lever draws the entire telescope towards the centre, thereby probably bending the telescope somewhat. Therefore, Mr Aubert has done right in fixing the lever RS at the level of the centre at y.

Mr Aubert's arrangement for moving his quadrant. ABCD is a quadrangular support and can be moved by 4 rollers at C and D etc. The movable axis EF supports GHIK. Everything is of wood except the screw E and the spigots GH. It is operated in the following way: When the axis is turned around the spigots G and H are screwed or lifted upwards so that they catch the two suspension plates of the quadrant, and the quadrant which has been lifted off, is now hanging on the vehicle.

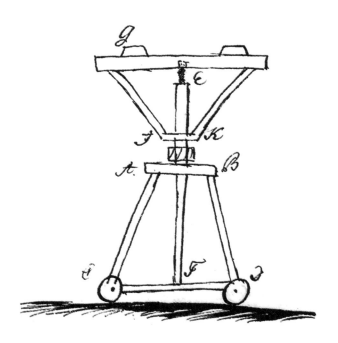

Thus two persons can easily transport it to the western side.

3°) There are two clocks: [one is] a regulator by Shelton. Regarding the second hand it is worth noting that every 5 seconds have been marked with a long mark in order to avoid miscounts.

I have already mentioned that the suspension of the clock is placed on the wall, and not on the case which has been fastened separately by means of screws.

The second clock is a counter which beats the seconds distinctly and loudly, but without striking the seconds. The minutes are marked with similar strokes of the clock.

4°) An equatorial support for a 3-foot reflector with objective micrometer. The support is a fixed tripod. A brass plate ab, fastened to a piece of wood, has been placed on the horizontal plate CD. Its angle with the plate CD is the same as the equatorial altitude. The brass plate de is parallel to it and has been divided into hours and minutes. gh is

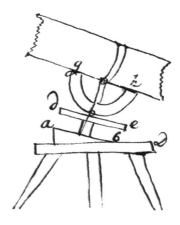

November 1777
London

the declination plate. It is easy to see that it is possible to seek out any celestial body and to follow its diurnal movement. Such an instrument can be used only in a certain or given place.

5. *An achromatic 3½-foot telescope* on a parallactic support of the same design as the one in the Oxford Observatory. Mr Aubert has improved it by placing a leaden ring, covered with cloth, around the telescope, in order to balance everything.

6. *An equatorial support* which he could mount on the window shutter after having secured it carefully.[242]

7. A very comfortable kind of small chairs which are said to have been invented by Smeaton.[243] ABDE is a round chair with three or four legs. It is about 1 to 1½ foot high. FHIG is a quadrangular, hollow tube fastened to it. KL is a round disc, movable by means of a handle.

242 The text in the drawing seems to be vindus skodder = window frame.

243 The civil engineer and inventor John Smeaton (1724–1792).

Through it is a hole with a screw-thread. MN is the seat, about 7 inches in diameter, to which the screw QP has been fixed. It is then obvious that when the disc KL is turned round, it is possible to raise or lower the seat at will until you are comfortably seated at the transit instrument or quadrant.

It would be very comfortable with a few of these astronomical chairs of various heights.

8. A small, movable quadrant of the same design as the large 3-foot quadrant at Greenwich.

9. Furthermore, Mr Aubert had a considerable collection of other instruments relating to astronomy, such as Hadley's sextants and octants of

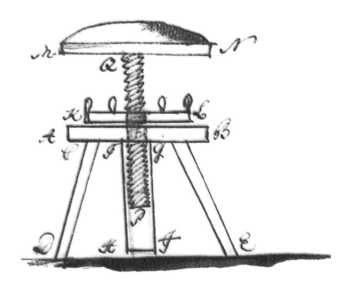

November 1777
London

Note. Dollond has attached
such a device to a hand-held
microscope in order to focus
the object.

various sizes; globes, celestial maps, orreries, terrestrial tele-scopes etc.

The latter were fitted with a device for moving the ocular by a screw or a wheel and a serrated plate as shown in the figure.

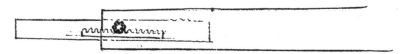

Mr Aubert's observatory is the most complete in Europe for its size. All the instruments were made by Bird. The clocks follow sidereal time. Mr Aubert himself is a very kind and shrewd man. At present he is a merchant, but he considers retiring one day and devoting himself entirely to astronomy.

In my description of the Greenwich Observatory[244] I will re-fer to the drawing shown under October 18th.

The ground floor of A is Doctor

244 For details on the instruments, see Howse 1975, Howse 1986, pp. 72–73 and Howse 1994, p. 215.

Maskelyne's residence. The top floor is a large octagonal room with a view over the clear horizon. In this room all kinds of telescopes are stored. The most remarkable among them were a) a 6-foot reflecting telescope by Short, b) a 3-foot reflecting telescope with an objective micrometer, c) a 10-foot Dollond with two glasses, d) a 3½-foot Dollond with three glasses.

Some of these have several different devices in the supports. The 10-foot Dollond was mounted as shown in the figure.

The front part is hung up in a pulley B attached to the support AB and counterbalanced by means of the weight P. The back part C is supported by a footing D, whose bar is threaded. It can be raised or lowered in the same way

November 1777
London

as shown in my description of the astronomical chair.

Several other telescopes have a parallactic support adapted for Greenwich polar altitude.

A device for diminishing the light by means of two square plates moving above each other. For that purpose Doctor Maskelyne had a curious device, not yet completely finished.

In the small buildings or towers B and C there are two equatorial sectors made by Bird; however, they are not quite satisfactory. Ramsden has been given the task of constructing a new equatorial instrument; he has told me about some of his ideas on that subject, but this instrument is not likely to be finished within the next few years. The construction of the present equatorial sector in Greenwich is the same as that of the equatorial sector in Oxford; except that the dimensions are here somewhat larger.

The building DFE is the real observatory. In E the transit instrument is placed. It supports an 8-foot achromatic telescope. In order to eliminate the influence of the solar rays on the tube, Doctor Maskelyne had covered it with white paper. I am more inclined to think that this would have a bad effect and increase the changes of the tube by heating, because the white paper is pasted immediately on to the brass. On the contrary, if it is held as a screen in front of the tube without touching it, it will no doubt have a good effect.

Furthermore, the following instruments are placed in E.

1) a clock by Graham with a gridiron pendulum. Arnold has made the pallets and the bearing[245] from rubies and diamonds. He said that the former ones made of steel were found to be worn out. And Doctor Maskelyne maintained that the clock goes much better now than before.

245 Our suggested translation for the word used by Bugge here, 'formgirer', which is not known.

Novemb. 1777
London

2) a new sea clock by Mudge.[246] It was round with a diameter of about 5 inches, and a height of 3 inches. Doctor Maskelyne had tested it thoroughly for 3 to 4 weeks and he found that it worked very well. It is wound up every day, but it can go for 48 hours.

3) A 12-foot sector by Graham. It is placed beside the wall. It has probably been used by Bradley and it is described in Smith's Optics.

F or the Middle Room is a small astronomical library, and the living-room for the observers.

D is the room where the two 8-foot mural quadrants are placed. The first, completely made of brass, was constructed by Bird, according to the methods described in his treatise on the construction of mural quadrants. It faces south. The second mural quadrant was originally constructed by Graham, but later on Bird gave it a new graduation. Only the edge is made of brass,

246 This was the first marine timekeeper made by the horologist Thomas Mudge (1715/16–1794) illustrated opposite. Based on tests at Greenwich in the years 1774–1778, Maskelyne said that it was 'greatly Superior in point of accuracy to any timekeeper which hath come under my inspection'. Bugge's 'round' may be confusing, as it is in an octagonal wooden box, but it is clear from his drawing that it was not one of the pair of marine timekeepers that Mudge made in 1777, known as 'Blue' and 'Green'. See *Your Time* (2008), pp. 10–14.

▲ *Marine timekeeper made by Thomas Mudge, London, c. 1774. Diameter 13.25 cm. Mudge explained that he had made the case in wood rather than in brass 'to save money, of which I have had, at no time, much to spare'. British Museum, inv. CAI-2119.*

▲ Harrison's First Marine Timekeeper H1, completed in 1735 and tested at sea the following year. Height 67 cm. Royal Observatory, Greenwich (National Maritime Museum). This massive contraption was intended to be unaffected by the motion of a ship owing to its two interconnected swinging balances and compensates for changes in temperature.

▲ Harrison's Second Marine Timekeeper H2, made between 1737 and 1739. Height 68.5 cm. Royal Observatory, Greenwich (National Maritime Museum). Larger and more solidly built than H1, and with additional refinements, it was never tested at sea. Harrison kept it running at his house for many years until, in 1766, it was taken from him by the Astronomer Royal under the conditions of the longitude prize.

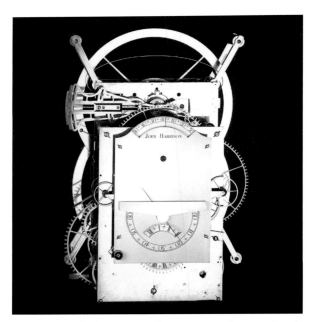

▲ Harrison's Third Marine Timekeeper, H3. Height 62 cm. Royal Observatory, Greenwich (National Maritime Museum). Started in 1740, it took Harrison nearly 19 years to build and adjust, although it was not to win him the great longitude prize.

▲ Harrison's Fourth Marine Timekeeper, H4. Diameter 13 cm. Royal Observatory, Greenwich (National Maritime Museum). This is Harrison's prize-winning longitude watch, completed in 1759. With its very stable, high frequency balance, it proved the successful design.

and the rest is made of iron. This quadrant faces north. In this room there were several instruments by the elder Harrison: four devices for the invention of the sea clock. They were large and very composite.[247]

G is a building with a very nice movable quadrant with a radius of 3 feet. The support is, in principle, the same as that of the big geographical instrument which I have employed for surveying Zeeland [Sjælland],[248] or as that of the new quadrant now under construction by Mr Ahl in Copenhagen for the observatory.[249] Therefore I will not waste my time over a detailed description. The structure of the quadrant was almost as shown in the above figure. It has a spirit level as well as a weight, and the telescope is adjustable only to the various observable altitudes.

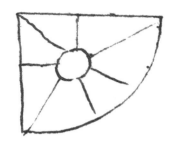

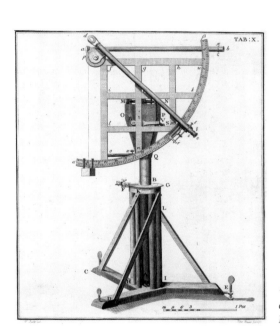

◀ *Engraving of Ahl's quadrant 1777 from Bugge 1784, Tab. X.*

▲ *Ahl's quadrant now in the Museum of the Round Tower, Copenhagen*

247 These were the celebrated H1–4 (illustrated opposite), made by the Lincolnshire carpenter and clock maker John Harrison (1693–1776) in his, ultimately succesful attempt to solve the biggest scientific problem of the 17th and 18th centuries: the navigational problem of finding the longitude at sea. One solution was known to be the use of accurate, robust clocks on ships at sea, keeping 'home time' to compare with 'local time' to give one's position east or west of the home port.

248 On this instrument, see fol. 76 recto.

249 The 3-foot portable quadrant, with which Bugge determined the latitude of the Round Tower observatory in the winter of 1778–79, survives (but missing the telescope) in the Round Tower Museum, see illustrations. Evidently, Ahl had begun its construction before Bugge left Copenhagen.

Novemb. 1777
London

Doctor Maskelyne thought to have observed, by means of a spirit level, that repeated reversals of the telescope gave rise to inaccuracies. Therefore he had constructed the following *plumb spirit level* in order to check the horizontal position of the axis.

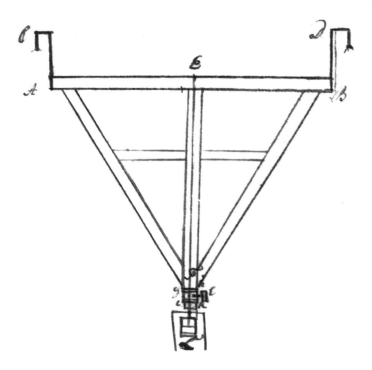

C and D are the two usual hooks which were attached to the axis of the transit instrument. AEBF is a mahogany triangle with its requisite connections, to the back of which transverse rulers have been attached. gikh is a brass plate which, by means of a very fine screw l, can be moved backwards and forwards in its edges. The pitch of this screw has a certain proportion to the inch, and the circumference in l has again been divided into certain parts in the proportion of 1 to 60. Consequently it is possible to know very exactly how many thousandths of an inch and how many seconds (in relation to the radius EF) the plate has moved. In the brass plate gikh is a very small hole which is illuminated by a candle, standing behind it. Its construction will be described later on. From E the plumb L is suspended by a fine metal thread. Thus the thread can very easily cut the illuminated hole, giving the same intensity of light on both sides.

EF is about 4 feet

Novemb. 1777
London

The device supporting the water pot and the candle is as follows:

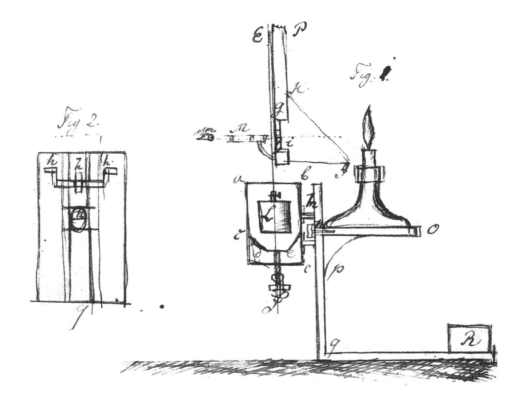

Pgi is the central ruler. EL is the thread and the plumb. gi is the brass plate. M is a microscope. hnpqR is a right-angled wooden support. On this (as shown in figure 2) the support pno can be carried up and down, so that the light has the correct height with respect to the hole in the brass plate. Let me add that a piece of waxed paper IK will make the shine of the light more even and comfortable.

At h there is a piece of brass, to which, by means of an edge h, the pitcher abcde can be fastened, and in which the plumb is floating. This pitcher has a conical bottom cde, so constructed that the plumb will always remain in the centre. When the plumb is not to be used, the bottom is screwed up by means of the screw F in order to remove the plumb L from the metal thread. Besides there is another screw in the bottom. When it is removed, the water can be drained off. R is a weight which balances the whole device, so that it does not overturn.

The verification of the device must naturally take place by means of reversals. Half of the error is corrected by the screw C on the perpendicular, and the other half of the error must be eliminated by raising or lowering the axis of the transit instrument.

Doctor Maskelyne verified it in my presence, and it was found that after the reversal the hole was cut in the centre by the thread.

The spirit level in the Greenwich Observatory has been made by Nairne with much care; and he himself has cut the interior of the glass tube to be cylindrical. I would think that the experiments should be carried out with much accuracy and care before a good spirit level is put aside and replaced by a weight of 4 feet.

The Greenwich clocks follow sidereal time.

On my last day in London I saw the instruments which Mr Nairne and Blunt had made for me,[250] and I found them all very pleasing; they were:

---

250 See appendix 3 'Instruments bought in London'.

1°) an air pump with accessories for all necessary experiments. 2°) An electrical machine with equipment, and a battery with 9 glasses. 3°) Models of the simple machines. 4°) a hydrostatic balance. 5°) Concave, convex, and cylindrical mirrors complete with drawings. 6°) achromatic prisms. 7°) a thermometer with Fahrenheit and Reaumur graduations. 8°) two different kinds of the finest metal threads for astronomical use. Total price about 600 rigsdaler.

On November 10th I left London and went via Osnabruck to Hamburg and from there to Copenhagen where I arrived safely on December 1st, 1777.

Bugge[251]

[fols. 92 verso – 94 recto are blank. On fol. 94 verso are some calculations]

251 The text from 'Total price' onward is written in much darker ink and clearly
       a later entry.

Appendix 1

TRAVEL EXPENSES

Together with the five notebooks of the journal is a separate document of two sheets, on which Bugge wrote down his travel expenses for the first weeks until his arrival in London. It is transcribed here. Whether there ever was a similar document for the later parts of his travels is not known.

The monetary standard in Denmark in Bugge's time was

1 rigsdaler (rdr) = 6 mark
1 mark   = 16 skilling
1 skilling   = 3 hvid

On August 2nd I went from Copenhagen to Roskilde.
August 3rd from Roskilde to Korsør.
August 4th crossed the Great Belt to Nyborg.
August 6th from Nyborg to Aaresund.
The 7th reached halfway between Flensburg and Schleswig.
The 8th to Itzehoe.
The 9th to Hamburg.

August 1777. The journey from Copenhagen to Hamburg.

Travel expenses from Copenhagen to Hamburg

received	90	rdr.	
To Hamburg spent	47	-	2
Remainder	42	-	4
Took in Altona	100		
Summa	142	-	4

Bought in Hamburg a purse	1	-		
2 pairs of silk stockings	6	-		
1 hat	3	-		
Trousers	4	-	4	6
Tailor's wages	1	-		
1 brush	-	-	2	
1 cane	2	-		
Black stockings	2	-		
Pocket money	4	-		
Lodgings	7	-		
Hired servant for 3 M[ark] per day	4	-		
Post bill	1	-		

In Hamburg spent	36	rdr.
Remainder	106	rdr.	4

Remainder upon departure from Hamburg	106 rdr.	4

Coach fare from Hamburg to Bremen	3	rdr.
Coach fare from Bremen to Leer	3	rdr.
From Leer to Nieuweschans	1	rdr.
From Nieuweschans to Groningen	1	rdr.
From Groningen to Lemmer	3	rdr.
Lemmer to Amsterdam via Zuider Zee	1	rdr.
Food from … to …	8	rdr.

	20	rdr.

Remainder upon arrival in Amsterdam	86 rdr.	4
Received from Mr Hasselgreen	100 rdr.	2

Summa	187	rdr.

Spent in Amsterdam

Books	5	rdr.
Maps	4	rdr.
Expenses to see various things	6	rdr.
The journey to Zaandam	4	rdr.
Haircut and shave	3	rdr.
Food for 9 days	9	rdr.
Gratuities	1	rdr.
Hired servant à 30 stivers	6	rdr.

	38	rdr.

Remainder in Amsterdam	149	rdr.

Remainder upon departure from Amsterdam	149	rdr.
The journey from Amsterdam to Leiden	1	rdr.
Paid at Cour d'Holland for 3 Nights and 2 days	4	rdr.
To see the Academy, the garden, and the cabinets	1	rdr.
The journey from Leiden to The Hague	1	rdr.
Food in The Hague	4	rdr.
To see the Prince's cabinet and library	2	rdr.
Travel expenses from The Hague to Rotterdam	1	rdr.
Food in Rotterdam	2	rdr.
Paid for a hired servant for 6 days and his return journey	4	rdr.
The journey from Rotterdam to Hellevoetsluis	2	rdr.
Food in Hellevoetsluis with ship provisions	2	rdr.
Paid for Packet boat to Cap[tain]. 5 rdr. to ship crew 1 rdr.	6	rdr.
Custom house in Harwich	2	rdr.
The journey from Harwich to London	6	rdr.
Food	1	rdr.
	38	rdr.
Remainder upon arrival in London	111	rdr.

I had this money in Dutch Ducats and will keep it until my return journey. Since I received

90 rdr. and from the merchants received 200 rdr. it follows that the whole journey from home through Holland until my arrival in London cost in total 179 rdr.

Appendix 2

BOOKS BOUGHT IN LONDON

On fol. 55 recto – fol. 56 verso, Bugge lists three dozen books bought during his stay in London, mostly from John Nourse's bookshop in the Strand, the street that links Westminster to the City. Total expenditure on books was £ 34 6s 6d.

Practically all titles could be identified using the 'advanced search' facility in the on-line catalogue of the British Library. In our notes we copy the British Library descriptions.

1. Books bought in London

	£	shilling
Martin's Philosophia Brittan[1]	-	18
… achromatic Optics[2]	-	6
… Experimental Philos[3]		3 ½
Dunn's practical Astronomy[4]	-	12
Fergusons Works[5]	1	18
Hevelii Selenographia[6]	1	1
Ludlams Observations[7]	-	11
Dunn's Variations Atlas[8]	1	12
Cumming Clock-Theory[9]	-	18
Total	8	19 ½

At Nourse's in the Strand[10] I bought the following:

Beccaria on Electricity[11]	1	1
Robertsons Navigation[12]	-	18
Robins Tract[13]	-	10 ½
Simpsons Fluxions[14]	-	12
– Exercises[15]	-	5 ½
– Annuities[16]	-	3
– Essays[17]	-	6
– Dissertations[18]	-	7
– Tracts[19]	-	7
Dodsons Repository[20]	-	12
Landens Lucubrations[21]	-	6
Harrisons Time Keeper[22]	-	5
Maskelynes of its going[23]	-	2 ½
Bird Dividing and Construction of mural Qvadrants[24]	-	2 ½
Adjutary Tables to Nautical Almanac[25]	-	2 ½
Maskelynes Observations[26]	1	11
Phipps' Voyages[27]	-	12 ½

Refractions and Parallax tables[28]	1	16
Hollidays Fluxions[29]	-	6
Ludlam on Hadleys Octant[30]	-	4 ½
Astronomical Observations in the South Sea[31]	1	1
Costards History of Astron.[32]	-	10 ½
Jones Geography[33]	-	3 ½
de la Lande Ephemerides for 1775 to 1785[34]	1	1
Summa	13 £	6 ½ sh
Earlier	8 £	19 ½
Bailey Machines including prepaid subscription 2 Vol.[35]	4	4
Summa. A.	26 £	10 A
Also from Nourse		
Emerson Works	4	4
– Mechanics[36]		15
– Increments[37]	-	7 ½
Nautical Almanak from the beginning to now	2	5
Ramsdens Engine for Dividing mathematical Instruments[38]	-	5
Summa. B.	7 £	16 ½ B

Total amount paid for books
Total A   26 £   10 sh
Total B   7 £   16 ½

Grand total   34 £   6½ sh.

1   Benjamin Martin, *Philosophia Britannica; or, a new … system of the Newtonian Philosophy, Astronomy and Geography. In a course of twelve lectures, with notes, containing the … proofs … of all the principal propositions in … Natural Science … with … copper-plates*. Third edition. 3 vol. London, 1771. 80.

2   Probably Benjamin Martin, *New elements of optics; or, The theory of the aberrations, dissipation, and colours of light: of the general and specific refractive powers and densities of mediums; the properties of single and compound lenses: and the nature, construction, and use of refracting and reflecting telescopes and microscopes, etc.* London: the author, 1759. 2 pt.: pl. VIII. 80.

3   Benjamin Martin, *A course of lectures in natural and experimental philosophy, geography and astronomy: in which the properties affections and phænomena of natural bodies … are … explain'd on the principles of the Newtonian philosophy, etc.* Reading: printed, and sold by J. Newbery & C. Micklewright, 1743. pp. 126: pl. VIII. 40.

4   Samuel Dunn, *A new and general introduction to practical Astronomy: with its application to Geography. … Topography, etc.* London, 1774. 80

5   It is not clear which publication by the lecturer of natural philosophy and inventor of scientific instruments, James Ferguson (1710-1776), Bugge bought; as far as we know the 5-volume edition of his Works did not appear until 1823.

6   [Johannes Hevelius], *J. Hevelii Selenographia: sive, Lunæ Descriptio. Addita est, Lentes expoliendi Nova Ratio; ut et Telescopia diversa construendi.* Gdansk, 1647. fol.

7   William Ludlam, *Astronomical Observations, made in St. John's College, Cambridge, in the years 1767 and 1768; with an account of several Astronomical Instruments.* Cambridge, 1769. 40

8   Samuel Dunn, *A new atlas of the mundane system; or, of Geography and Cosmography: describing the heavens and the earth, etc.* London, 1774. fol.

9   Alexander Cumming, F.R.S., *The elements of clock and watch-work, adapted to practice. In two essays.* London: printed for the author, 1766. 40

10 Listed as 'Nourse, Bookseller, 138, Strand', in Pendred 1785. John Nourse was one of the principal importers of French and Continental books, and one of the book-sellers who travelled widely on business; see Barber 1975, pp. 234-5.

11 Giambattista Beccaria, *A Treatise upon Artificial Electricity, in which are given solutions of a number of electric phoenomena, hitherto unexplained. To which is added, an Essay on the mild and slow electricity which prevails in the atmosphere during serene weather; translated from the Italian, etc.* pp. iv. 457. pl. XI. J. Nourse: London, 1776. 40

12 John Robertson, *The Elements of Navigation, containing the theory and practice, with the necessary tables. To which is added a treatise on Marine Fortification … Third edition, with additions, and compendiums for finding the latitude and longitude at sea.* 2 vol. London, 1772. 80

13 Benjamin Robins, *Mathematical Tracts … Published by J. Wilson* (London, 2 vols, 1761)

14 Thomas Simpson, F.R.S., *The Doctrine and Application of Fluxions. Containing … a number of new improvements in the theory. And the solution of a variety of … problems in … the mathematicks. Second edition, revised, etc.* 2 pt. London, 1776. 80

15 Thomas Simpson, F.R.S., *Select Exercises for young proficients in the Mathematicks, etc.* London, 1752. 80

16 Thomas Simpson, F.R.S., *The Doctrine of Annuities and Reversions, … with … tables, etc.* London, 1742. 80

17 Thomas Simpson, F.R.S., *Essays on several curious and useful subjects, in speculative and mix'd Mathematicks, etc.* London, 1740. 40

18 Thomas Simpson, F.R.S., *Mathematical Dissertations on a variety of physical and analytical subjects, etc.* London, 1743. 40

19 Thomas Simpson, F.R.S., *Miscellaneous Tracts on some curious ... subjects in Mechanics, Physical Astronomy, and Speculative Mathematics; wherein the precession of the equinox, the nutation of the earth's axis, and the motion of the moon in her orbit, are determined.* London, 1757. 40

20 James Dodson, *The Mathematical Repository* ... The second edition. pp. xi. 336. J. Nourse: London, 1775. 120

21 John Landen, *Mathematical Lucubrations; containing new improvements in various branches of the Mathematics.* London. 1755. 40

22 John Harrison, *Principles of Mr. Harrison's Time-keeper, with plates of the same; published by order of the Commissioners of Longitude.* pp. 31. J. Nourse: London, 1767. 40

23 Nevil Maskelyne, *The original observations of the going of the watch [i.e. Mr Harrison's] from day to day. (Comparisons of Mr Harrison's watch with the mean time ... Appendix containing observations of equal altitudes of the sun ... according to the time of Mr Harrison's watch, etc.)* [London, 1768?] 40

24 Evidently bound together two tracts by John Bird, *The Method of Dividing Astronomical Instruments pp. vi, 14. John Nourse: London 1767. 40 and The Method of Constructing Mural Quadrants, exemplified by a description of the brass mural quadrant in the Royal Observatory at Greenwich.* pp. 27, pl. IV. John Nourse: London, 1768. 40

25 Probably first or second edition of: *Tables requisite to be used with the Nautical Ephemeris ... The third edition ... improved. (The explanation and use of the tables [by W. Wales]. Appendix to the third edition ...; being new tables of natural sines, natural versed sines, and logarithms of numbers.)* [Edited by N. Maskelyne.] ENGLAND. Departments of State and Official Bodies. Commissioners of Longitude. 3 pt. London, 1802. 80.

26 Not specific enough to determine which of Maskelyne's published astronomical observations Bugge bought.

27 Constantine John Phipps, 2nd Baron Mulgrave, *The Journal of a Voyage undertaken by order of his present Majesty, for making discoveries towards the North Pole, by the Hon. Commodore Phipps, and Captain Lutwidge, in his Majesty's sloops Racehorse and Carcase. To which is prefixed, An Account of the several Voyages undertaken for the discovery of a North East Passage to China and Japan.* pp. xxviii. 118. F. Newbery: London, 1774. 80.

28 *Tables for correcting the apparent Distance of the Moon and a Star from the effects of Refraction and Parallax. Published by order of the Commissioners.* [Edited by A. Shepherd.] ENGLAND. Departments of State and Official Bodies. Commissioners of Longitude. J. Archdeacon: Cambridge, 1772. 40.

29 Francis Holliday, *An Introduction to Fluxions.* London, 1777. 80

30 Either William Ludlam, *Directions for the use of Hadley's Quadrant; with remarks on the construction of that instrument.* pp. ix. 137. London: printed by R. Hall. Sold by T. Cadell, 1771. plates. 80., or his *The theory of Hadley's quadrant, or The rules for the construction and use of that instrument demonstrated.* London: printed by R. Hett; sold by T. Cadell, 1771. pp. 30: plates. 80.

31 Possibly = *The Original Astronomical Observations, made in the course of a Voyage towards the South Pole, and round the World, in his Majesty's Ships the Resolution and Adventure, in the years MDCCLXXII, MDCCLXXIII, MDCCLXXIV, and MDCCLXXV, by William Wales ... and Mr. William Bayly, etc.* [Edited by W. Wales.] ENGLAND. Departments of State and Official Bodies. Commissioners of Longitude. pp. lv. 385. pl. II. J. Nourse; J. Mount & T. Page: London, 1777. 40

32   George Costard, *The History of Astronomy, with its application to geography, history, and chronology; occasionally exemplified by the globes.* pp. xvi. 308. J. Newbery: London, 1767. 40

33   Possibly Evan Jones, *The Young Geographer and Astronomer's Best Companion* (London 1773)

34   Stellar lists, calculated and published by the French astronomer Joseph Jérôme Le Français de Lalande (1732-1807).

35   For Bailey and the book that Bugge bought from him, see Bugge's journal fol. 44 recto.

36   William Emerson, *Mechanics; or, The doctrine of motion, etc.* [The preface signed: W. Emerson.] London: J. Nourse, 1769. pp. iv, 148: pl. IX. 80

37   William Emerson, *The Method of Increments, wherein the principles are demonstrated, and the practice thereof shewn in the solution of problems.* London, 1763. 40

38   Jesse Ramsden, *Description of an engine for dividing mathematical instruments. ... Published by order of the Commissioners of Longitude.* pp. 14. J. Nourse: London, 1777. 40

Appendix 3

INSTRUMENTS BOUGHT IN LONDON

On fol. 56 verso – fol. 57 verso, Bugge lists some fifty instruments, bought during his stay in London. Apart from a few supplied by Peter and John Dollond, George Adams and Addison Smith, the bulk was a set of apparatus for natural philosophy experiments from the workshop operated jointly by Edward Nairne and Thomas Blunt. Total expenditure on instruments was £ 88 4s.

When he visited the workshop on 15 September, Bugge noted that he had been given a catalogue of its products, and presumably he then ordered the set. He returned to the workshop eight weeks later, on his last day in London, to inspect the instruments, and "found them all very pleasing". They would then have been dispatched to Copenhagen.

Bugge wrote the list of Nairne and Blunt equipment in English, and incidentally in a different ink from the preceding entries. We have compared it with the only known surviving copy of a printed catalogue of the firm, a 6-page *Prices of some of the optical, mathematical and philosophical instruments / made and sold by Nairne and Blunt, no. 20 in Cornhill, opposite the Royal Exchange, London,* tentatively dated 1780.[1] Although each item can somehow be matched, and prices do correspond (he seems to have gone for mid-price range), Bugge's list is not a verbatim transcription of the relevant entries in that catalogue. The item "4 Packing Cases for above ment[ioned] Inst[ruments]" suggests that he copied the workshop's confirmation of his order, or the invoice.

---

1   It is in the Harvard Science Centre, Harvard University, Cambridge, Ma., USA. We are grateful to Sara Schechner for supplying a scan of this unique document.

2. Instruments bought in London	£	shilling	(d = pence)

		£	shilling	
St. Paul's Church	Dollond's compound microscope	8	8	
Hay Market	Dollond's new parallel ruler	2		
	Prismata achromatica at Smith's	2		
	Rule by Smith with French, English, Dutch and Antwerp foot measures	-	3	
	A.) Total	12	11	
	A mathematical set of drawing instruments bought at Adams'	2	2	
	Crown glass and flint glass bought from Smith	8	9	
At Mr Nairne & Blunt's	Hydrostatic Balance	7	17	6d
	An Electrical Machine	13	13	
	A Battery of 9 Jars	4	4	
	The mechanic Powers	21		
	Thermometer	2	2	
	Two Glasses for a Hadley's Quadrant	-	3	6d

	£	shilling	(d = pence)
A double barrelled Air Pump	11	15	
A Hand glas	-	1	
4 Cylinder Glasses	-	8	
3 Straight Receivers	-	7	
2 Bell Glasses	-	6	
A Guinea and Feather Glass with Instrument	2	2	6d
Bladder Glas		2	6d
A Lung Glass	-	5	
A Bras[s] Cone	-	3	6d
Small Glass Fountains	-	3	6d
Bell and Apparatus	-	13	6d
Apparatus of Hemispheres	1	1	
A double Transferer	2	10	
A Bladder	-	2	6d
Large Receivers for Animals	-	12	
One Dozen Sqvare Vials	-	6	
A Square Vials cap'd with Valve	-	1	6d
Cage	-	3	
Receiver Box plate and Lead Weights	-	15	
Cork and Lead	-	-	6d
2 Blank Screws	-	2	
Copper Pipe Plate and Stop Cock	-	12	
Syringe and Lead	-	13	6d
3 Bubles and 3 Bolt Heads	-	2	
Receiver for Torricellian Experiments	-	12	

	£	shilling	(d = pence)
A 4 Inch Plate with Collar and Leader	-	5	
A Florence Flask² with Balance Weight	-	3	6d
A Scale Beam	-	11	
A Glass Tube and Receiver. Barom. Exp.	-	15	
A Receiver with Brass Plate at Top with Cock and small Receiver under the large one	1	10	
A Receiver for Shower of Mercury	-	10	
A Receiver with a Piece of Cemented Cane	-	2	
A Receiver with Barometer Tube	-	10	6d
A brass Ball with Centrep. Scale Beam and Receiv.	1	1	
A small exhausting syringe	-	15	
8 pounds Cast Steel Barrs	-	9	4d
8 pounds Forged Ditto	-	4	8d
4 Packing Cases for above ment. Inst.	1	9	
A half Cylinder Mirror and 6 Prints	2	2	
A 16[?] inch Concave and convexe Mirror in black frame	3	3	
A 8 inch Cylinder Ditto	1	11	6d
Total	88	4	

[follow subtractions with 49 and 12]

---

2   Oxford English Dictionary: "a flask of the kind used to contain Florence oil; a similar vessel for use in a laboratory".

Appendix 4

DATA OF ASTRONOMICAL OBSERVATIONS

On 9th October, during Bugge's visit to the Radcliffe Observatory in Oxford, professor Hornsby told his visitor about his work on the proper motion of the star Arcturus. He had found suspected misprints in Horrebow's publication of Ole Roemer's observations of the same star. He therefore requested that Bugge, upon his return to Copenhagen, would compare them with Roemer's manuscript.[1] Three days later, Bugge returned to this matter in his journal and wrote down the relevant astronomical data, transcribed below.

In 1779 Bugge delivered an address at the anniversary celebration of Copenhagen University about astronomical observations and observatories of the past, which ended with a detailed report on the English observatories that he had visited two years earlier. He stated that he had indeed found Hornsby's suspicions confirmed and had reported this to his Oxford colleague.[2]

Bugge also wrote down Hornsby's own observations of the eclipse of Jupiter's satellites for 1774 and 1775, and these too are transcribed below.

1   See fol. 62 verso.

2   Bugge 1779, p. 19.

Professor Hornsby has observed a peculiar movement of Arcturus which he has described in the Philosophical Transactions. He has continued these observations, and for comparison he then uses old observations among which Roemer's Triduum deserves special note. He gave me the following list of various places, where – in his opinion – the value of the minutes is incorrect:
Roemeri Triduum in Operibus P. Horrebow, October 20. Ceti Caudæ Lucida 0 h 28' 7 1/3. Correct. 0.29 7 1/3.

correspondentes altitudines Solis die 22 Octob.

		correct.
10 h	54.49 -	10 53.49 -
10	55.28 -	10 54.28 -
10	55.58 v	10 55.58 v

October 23. Sirii Auris Sinistra

		correct
6 --		6 --
6	51.58 v	6 50.58 v
6	52.22	6 51.22
6	52.47 v	6 51.47 v

In the Demonstratione Parallaxeos orbi annui are indicated some observations of Sirius, Lyra, Pes geminorum, Caput Draconis, Capella, for the years 1701,2,3,4,5,6,7,8,9. The question is: *Do these original observations still exist, and have still more stars been observed by Horrebow?*

Professor Hornsby gave me the following observations of the eclipse of Jupiter's satellites, observed by himself at Oxford.

			temp.	
1774	Octob	21	08.30.25 3/4	Im 1. 2 mag Doll
	Nov	13	10.48.43.2	Em 1.
		29	09.02.54.0	Em 1. non valde clarum

			temp.	
1774	Dec.	30	09.33.38.3	Imm iii
			11.01.45.1	Em
1775	Feb.	22	07.44.32.2	Em 1 exit cum impetu
	Mart.	17	08.06.19.3	Em 1
		23	08.04.13.8	Em 2
	Aug.	19	12.08.18.75	Im 2 aer vaporibus plenus
			14.26.00.4	Em
	Sept.	27	16.56.25.9	Em 2 clarum
		29	17.17.22.4	Im 1
	Octob.	01	11.46.17.25	Im 1
		04	17.14.02.7	Im 2 non valde clarum
		15	09.09.28.7	Im 2
		22	11.47.44.3	Im 2
		-	17.31.02.7	Im 1 dubio
		24	12.00.22.0	Im 1
	Nov.	02	08.22.58.1	Im 1
		09	10.16.32.2	Im 1
		16	08.56.17.7	Im 2
	Dec.	11	08.22.00.1	Em 2
		-	08.48.41.2	Em 1
		27	06.58.25.0	Em 1 coelum paulisper vaporosum

The tenth parts of the seconds are given, and time has been accurately reduced from sidereal time.

The observations for 1776 and 1777 have not yet been reduced.

Bibliography

Much information was found in Danish, German, Dutch and English biographical dictionaries, which have entries on dozens of persons mentioned in Bugge's journal. This is only explicitly referenced where the provenance of the information is not obvious.

Alder, Ken (2002). *The Measure of all Things. The Seven-Year Odyssey that Transformed the World* (London).

Allan, D.G.C. and Abbott, John L., eds. (1992). *The Virtuoso Tribe of Arts & Sciences. Studies in the Eighteenth-Century Work and Membership of the London Society of Arts* (Athens, Georgia and London).

Altick, Richard D. (1978). *The Shows of London* (Cambridge Mass.).

Andersen, Einar (1968). *Thomas Bugge. Et mindeskrift i anledning af 150 årsdagen for hans død 15. januari 1815* (Copenhagen).

Andrewes, William J. H., ed. (1996), *The Quest for Longitude* (Cambridge, Mass.).

Angerstein (2001). *R.R. Angerstein's Illustrated Travel Diary, 1753-1755. Industry in England and Wales from a Swedish perspective.* Translated by Torsten and Peter Berg (Science Museum, London).

Atterbury, Paul (1998). *The Thames. From the Source to the Sea* (London).

Aubert, T. 'Alex Aubert, F.R.S., astronome, 1730-1805', *Notes and Records of the Royal Society* 9, pp. 79-85.

Bailey, William (1772). *The Advancement of Arts, Manufactures and Commerce, or Descriptions of the Useful Machines and Models contained in the Repository of the Society for the Encouragement of Arts, Manufactures and Commerce* (London). This first volume was re-issued in 1776, "carefully corrected and revised by Alexander Mabyn Bailey". The second volume appeared in 1779, followed by a two-volume edition in 1782.

Bailly, Jean Sylvain (1766). 'Essai sur la théorie des satellites de Jupiter' in *Mémoires de l'Académie Royale des Sciences* (Paris), pp. 580-667.

Bailly, Jean Sylvain (1771). *Mémoire sur les inegalités de la lumière des satellites de Jupiter, sur la mesure de leurs diamètres, et sur un moyen aussi simple que commode de rende les observations comparables, en remédiant à la difference des vues et des lunettes* (Paris).

Barber, Giles (1975). 'Pendred abroad. A view of the late eighteenth-century book trade in Europe', in R.W. Hunt and others, eds., *Studies in the Book Trade. In honour of Graham Pollard* (Oxford), pp. 231-277.

Barlow, William (1740). 'An Account of the Analogy betwixt English Weights and Measures of Capacity', *Philosophical Transactions of the Royal Society* 41, pp. 457-459.

Bedini, Silvio (1971). 'The Tube of Long Vision', *Physis, Rivista internazionale di storia della scienza* XIII, pp. 147-204, esp. pp. 197-201.

Beer, Carel de, ed. (1991). *The Art of Gunfounding. The Casting of Bronze Cannon in the late 18th century* (Jean Boudriot Publications, Ashley Lodge, Rotherfield, East Sussex).

Bélidor, Bernard Forest de (1737-39), *Architecture hydraulique; ou l'art de conduire, d'élever et de ménager les eaux pour les différents besoins de la vie* (Paris).

Bennett, J.A. (1987). *The Divided Circle. A History of Instruments for Astronomy, Navigation and Surveying* (Oxford).

Bennett, J.A. (1993). 'Equipping the Radcliffe Observatory: Thomas Hornsby and his instrument-makers', in: R.G.W. Anderson, J.A. Bennett and W.F. Ryan, eds., *Making Instruments Count: Essays on Historical Scientific Instruments presented to Gerard L'Estrange Turner* (Aldershot), pp. 232-241.

Bennett, James A. (2002). 'Shopping for Instruments in Paris and London', pp. 370-95 in Pamela Smith and Paula Findlen, eds., *Merchants and Marvels: Commerce, Science, and Art in Early Modern Europe* (New York and London).

Bennett, J.A., Johnston, S.A., Simcock, A.V. (2000). *Solomon's House in Oxford. New Finds from the First Museum* (Oxford: Museum of the History of Science).

Bernoulli, Jean (1771). *Lettres astronomiques où l'on donne une idée de l'état actuel de l'astronomie pratique dans plusieurs villes de l'Europe* (Berlin).

Bold, John (2000). *Greenwich. An architectural history of the Royal Hospital for Seamen and the Queen's House* (New Haven and London).

Boorsma, P. (1968). *Duizend Zaanse molens* (Amsterdam).

Bos, H.J.M. (1968). *Mechanical Instruments in the Utrecht University Museum. Descriptive Catalogue* (Utrecht).

Bradley (1798). *Astronomical Observations*, edited by Thomas Hornsby (London).

Brink, Paul van den, and Werner, Jan, eds. (1989). *Gesneden en gedrukt in de Kalverstraat. De kaarten- en atlassendrukkers in Amsterdam tot in de 19e eeuw* (Utrecht).

Britten, Frederick James (1982). *Britten's Old Clocks and Watches and their Makers* (London, 9th edition).

Bugge, Thomas (1779a). *Beskrivelse over den Opmaalings Maade, som er brugt ved de danske geographiske Karter; med tilføiet trigonometrisk Karte over Siæland, etc* (Copenhagen). A German edition was published in Dresden in 1787 as *Beschreibung der Ausmessungs-Methode, welche bey den Dänischen geographischen Karten angewendet worden.*

Bugge, Thomas (1779b). Anonymous, untitled address about the history of astronomical observations and instruments, mainly in England, printed in the Copenhagen University Programme for 1779; for details see the introduction.

Bugge, Thomas (1784). *Observationes astronomicæ annis 1781, 1782, & 1783. Institutæ in Observatorio Regio Havniensi et cum tabulis astronomicis comparatæ* (Copenhagen).

Camilleri, S (2001). 'Folders for Foreign Coins. English Folding Coin Balances for Weighing Foreign Gold Coins', *Equilibrium* nr. 2, pp. 2561-2568.

Cassidy, David A. (1985). 'Metereology in Mannheim: the Palatine Metereology Society, 1780-1795', *Sudhoff's Archiv* 69, pp. 8-25.

Christensen, Dan (1993). 'Spying on scientific instruments. The career of Jesper Bidstrup', *Centaurus* 36, pp. 209-244.

Christensen, Dan (2001). 'English Instrument Makers observed by Predatory Danes', in Lützen, pp. 47-63.

Christensen, Dan (2009). *Naturens tankelæser. En biografi om Hans Christian Ørsted* (Copenhagen).

Clercq, Peter de (1988). 'Science at Court: the 18th century cabinet of scientific instruments and models of the Dutch Stadholders', *Annals of Science* 45, pp. 113-152.

Clercq, Peter de (1994), *Het Koperen Kabinet. Schatkamers van de wetenschap 1550-1950* (Leiden: Museum Boerhaave).

Clercq, Peter de (1997a). *At the sign of the Oriental Lamp. The Musschenbroek workshop in Leiden, 1660-1750* (Rotterdam).

Clercq, Peter de (1997b). *The Leiden Cabinet of Physics. A Descriptive Catalogue* (Leiden: Museum Boerhaave).

Clercq, Peter de (2005a). 'Een Deense astronoom op bezoek in Nederland en Engeland. Het reisjournaal van Thomas Bugge uit 1777', pp. 74-84 in *Koersvast. Vijf eeuwen navigatie op zee. Een bundel opstellen aangeboden aan Willem Mörzer Bruyns bij zijn afscheid van het Nederlands Scheepvaartmuseum Amsterdam in 2005* (Zaltbommel).

Clercq, Peter de (2005b). 'A Dutch Gentleman in London. Anthony George Eckhardt, F.R.S. (1740-1810) and instruments of his invention', *Bulletin of the Scientific Instrument Society* 84, pp. 10-17.

Clercq, Peter de (2007). 'Private Instrument Collections Sold at Auction in London in the late 18th Century. Part 1: Professional Practitioners and Gentlemen-Collectors', *Bulletin of the Scientific Instrument Society* 95, pp. 28-36.

Clercq, Peter de (2009). 'Private Instrument Collections Sold at Auction in London in the late 18th Century. Part 2: Instruments and Watchmakers', *Bulletin of the Scientific Instrument Society* 100, pp. 27-35.

Clifton, G. (1995). *Directory of British Scientific Instrument Makers 1550-1851* (London).

Collins, Philip R. (2002). *Barographs* (Trowbridge).

Crawforth, M.A. (1979). *Weighing coins:English folding gold balances of the 18th and 19th centuries* (London).

Crawforth, M.A. (1985). 'Evidence from Trade Cards for the Scientific Instrument Industry', *Annals of Science* 42, pp. 453-554.

Crosland, Maurice P. (1969). *Science in France in the Revolutionary Era. Described by Thomas Bugge, Danish Astronomer Royal and Member of the International Commission on the Metric System (1798-1799).* Edited with an introduction and commentary by Maurice P. Crosland, with extracts from other contemporary works. Published jointly by the Society for the History of Technology and the MIT Press (Cambridge, Mass. and London).

Cumming, Alexander, F.R.S. (1766). *The elements of clock and watch-work, adapted to practice. In two essays.* (London: printed for the author).

Dekker, E. (1985). *The Leiden Sphere. An outstanding seventeenth-century planetarium* (Leiden: Museum Boerhaave).

Desaguliers, J.T. (1744). *A course of experimental philosophy* (London, second edition).

Dodson, James (1742). *The Anti-logarithmic Canon. Being a table of numbers consisting of eleven places of figures, corresponding to all logarithms under 100000 ... with ... a short account of logarithms, etc.* (London).

Dodson, James (1747). *The Calculator, being ... tables for computation, adapted to science, business, and pleasure* (London).

Dollond (1771). *Description and uses of the new invented universal equatorial instrument, or, portable observatory. With the divided object-glass micrometer. Made by P. and J. Dollond, opticians to His Majesty, in St. Paul's Churchyard, London* (London, 16 pages, 4°; undated, British Library catalogue suggests c. 1771).

Donnelly, M.C. (1973). *A Short History of Observatories* (Oregon).

Dunn, Samuel (1778). *The Theory and practice of the longitude at sea ... with a general introduction to its astronomical and physical principles ...* (London).

Engel (1986). *Hendrik Engel's Alphabetical List of Dutch Zoological Cabinets and Menageries,* second enlarged edition (Amsterdam).

Fas, J.A. (1775). *Inleiding tot de kennisse en het gebruyk der oneindig kleinen, in twee afdeelingen, behelzende de grondbeginselen van de differentiaal en integraal rekening: met een byvoegsel over het verdeelen van een gebrooke functie* (Leiden).

Ferrner, Bengt (1956). *Resa i Europa. En astronom, industrispion och teaterhabitué genom Danmark, Tyskland, Holland, England, Frankrike och Italien, 1758-1762* (Uppsala).

Forbes, R.J., and others, eds. (1969-1976). *Martinus van Marum. Life and Work* (6 vols., Haarlem and Leiden).

Geikie, Sir Archibald (1917). *Annals of the Royal Society Club. The record of a London dining-club in the eighteenth and nineteenth centuries* (London).

Guest, Ivor (1991). *Dr John Radcliffe and his Trust* (London).

Gunnis, Rupert (1968). *Dictionary of British sculptors 1660-1851* (London, new revised edition).

Handlist (1990): R.G.W. Anderson, J.E. Burnett and B. Gee, *Handlist of Scientific Instrument-Makers' Trade Catalogues 1600-1914* (Edinburgh: Royal Museum of Scotland: National Museums of Scotland Information Series No.8).

Heilbron, J.L. (1979). *Electricity in the 17th and 18th centuries. A study of early modern physics* (Berkeley).

Helden, Anne van, and Gent, Rob van (1995). *The Huygens collection* (Leiden: Museum Boerhaave).

Herk, Gijsbert van, with Herman Kleibrink (1983) *De Leidse Sterrewacht: vier eeuwen wacht bij dag en bij nacht* (Zwolle).

Hogg, Brig. O.F.C. (1963). *The Royal Arsenal: its Background, Origin and Subsequent History* (London, 2 vols).

Hornsby, T. (1773). 'An Inquiry into the Quantity and Direction of the Motion of Arcturus', *Philosophical Transactions* vol. 63, pp. 93-125.

Hornsby, T. , ed. (1798). *Astronomical Observations made at the Royal Observatory at Greenwich, from the year 1750 to the year 1762, by J. Bradley, D.D., Astronomer Royal* (Oxford, 2 vols, 1798-1805, vol. 1 edited by Hornsby).

Horrebow, P. (1735). *Basis astronomiæ, sive Astronomiæ pars mechanica in qua describuntur Observatoria, atque instrumenta Roemeriana Danica* (Copenhagen).

Howse, Derek (1975). *Greenwich Observatory: the Royal Observatory at Greenwich and Herstmonceux, 1675-1975. Vol. 3, The buildings and instruments* (London),

Howse, Derek (1986). 'The Greenwich List of Observatories: A World List of Astronomical Observatories, Instruments and Clocks, 1670-1850', a special number of *Journal for the History of Astronomy* 17.

Howse, Derek (1994). 'The Greenwich List of Observatories: Amendment List No. 1', *Journal for the History of Astronomy* 25, pp. 207-218.

Jackson, Melvin H., and Carel de Beer (1974), *Eighteenth Century Gunfounding. The Verbruggens at the Royal Brass Foundry. A Chapter in the History of Technology* (Newton Abbot).

Jong, E. de (1991). 'Nature and Art. The Leiden Hortus as a 'musaeum', in L. Tjon Sie Fat and E. de Jong, eds. *The Authentic Garden, a Symposium on Gardens* (Leiden), pp. 37-60.

Kernkamp, G.W. (1910). 'Bengt Ferrner's dagboek van zijne reis door Nederland in 1759', in *Bijdragen en Mededelingen van het Historisch Genootschap* 31, pp. 314-509

Keulen, E.O. van, Mörzer Bruyns, W.F.J., Spits, E.K. (1989), ‚*In de Gekroonde Lootsman*‘. *Het kaarten-, boekuitgevers en instrumentmakershuis Van Keulen te Amsterdam, 1680-1885* (Utrecht).

King, Henry C. (1954). *The History of the Telescope* (London).

King, J.C.H. (1996). 'New Evidence for the Contents of the Leverian Museum', *Journal for the History of Collections* 8, pp. 167-186.

Klessmann, Eckart (1981). *Geschichte der Stadt Hamburg* (Hamburg).

Kragh, Helge, and Peter C. Kjærgaard, Henry Nielsen, Kristian Hvidtfelt Nielsen, eds. (2008). *Science in Denmark. A Thousand-Year History* (Aarhus).

Kristensen, Leif Kahl (2001). 'Wessel as a cartographer', in Lützen, pp. 81-98.

Krogt, P. van de (1984). *Old Globes in the Netherlands* (Utrecht).

Levere, T. (1973). 'Teyler's Museum', in *Martinus van Marum. Life and Work* (6 vols., Leiden), vol. IV, pp. 39-102.

Lalande, J.J. de (1771). *Astronomie* (Paris, second revised and enlarged edition).

Lalande, J.J. de (1980). *Journal d'un voyage en Angleterre 1763*, ed. Hélène Monod-Cassidy (Oxford: Studies on Voltaire and the Eighteenth Century no. 184)

Loonstra, Marten (1985). *Het húijs int bosch: het Koninklijk Paleis Huis ten Bosch historisch gezien. The Royal Palace Huis ten Bosch in a historical view* (Zutphen; bilingual).

Lützen, Jesper, ed. (2001). *Around Caspar Wessel and the Geometric Representation of Complex Numbers. Proceedings of the Wessel Symposium at the Royal Danish Academy of Sciences and Letters, August 11-15 1998, Copenhagen* (Copenhagen: The Royal Danish Academy of Sciences and Letters, Matematisk-fysiske Meddelelser 46:2).

McConnell, Anita (2005). 'Origins of the marine barometer', *Annals of Science* 62, pp. 83-101

McConnell, Anita and Brech, Alison (1999). 'Nathaniel and Edward Pigott, itinerant astronomers', *Notes & Records of the Royal Society* 53, pp. 309-318.

McConnell, Anita (2007). *Jesse Ramden (1735-1800): London's Leading Scientific Instrument Maker* (Aldershot).

MacGregor, Arthur (2001). *The Ashmolean Museum. A brief history of the Museum and its collections* (Oxford).

Martin, Benjamin (1771). *Philosophia Britannica; or, a new system of the Newtonian Philosophy, Astronomy and Geography* (London, 3rd ed., 3 vols).

Maskelyne, Nevil (1771). 'Description of a Method of Measuring Differences of Right Ascension and Declination, with Dollond's Micrometer, together with Other New Applications of the Same', *Philosophical Transactions of the Royal Society* 61, pp. 536-546.

Maskelyne, Nevil (1776). *Astronomical Observations made at the Royal Observatory at Greenwich ... 1765 to 1774* (London), vol. 1, pp. i-vi.

Mazel, H. (1909). 'Van een aap in 1777', *Jaarboek Die Haghe*, pp. 361-380.

Mercer, Vaudrey (1972). *John Arnold & Son, Chronometer Makers, 1762-1843* (London).

Middleton, W.E. Knowles (1964). *The History of the Barometer* (Baltimore).

Millburn, John R. (1976). *Benjamin Martin. Author, Instrument-Maker and 'Country Showman'* (Leiden).

Millburn, John R. (2000). *Adams of Fleet Street. Instrument Makers to King George III* (Aldershot).

Mörzer Bruyns, W.F.J. (2003). *Schip Recht door Zee. De octant in de Republiek in de achttiende eeuw* (Amsterdam).

Molbech, C. (1843): *Det kongelige Danske Videnskabernes Selskabs Historie i dets første Aarhundrede 1742-1842* (Kjøbenhavn).

Morrison-Low, Alison (2007). *Making Scientific Instruments in the Industrial Revolution* (Aldershot).

Morton, A.Q., and Wess, J.A. (1993). *Public and Private Science: the King George III Collection at the Science Museum* (London and Oxford).

Mudge, John (1777). 'Directions for making the best composition for the metals for reflecting telescopes, together with a description of the process for grinding, polishing, and giving the great speculum the true parabolic curve', *Philosophical Transactions of the Royal Society* 67, pp. 296-349.

Nicolson, Benedict (1972). *The Treasures of the Foundling Hospital* (Oxford).

Nielsen, Keld (1989). *Hvordan Danmarkskortet kom til at ligne Danmark. Vedenskabernes Selskab opmåling 1762-1820* (Aarhus University, Videnskabshistorisk Museum).

Pedersen, Kurt Møller (1982). 'Thomas Bugge's dagbog 1777', in *Bibliotek for Læger* jrg. 174 Supplement 1, pp. 151-164.

Pedersen, Kurt Møller (2001). 'Thomas Bugge's Journal of a Voyage through Germany, Holland and England, 1777', in Lützen, pp. 29-46.

Pedersen, Olaf (1992). *Lovers of Learning. A History of the Royal Danish Academy of Sciences and Letters, 1742-1992* (Copenhagen)

Pendred (1785). *The Earliest Directory of the Book Trade by John Pendred (1785)*, ed. Graham Pollard (London 1955).

Peter-Raup, Hanna (1980). *Die Ikonographie des Oranjezaal* (Hildesheim/New York).

Phipps, Constantine John (1774). *A Voyage towards the North Pole undertaken by his Majesty's Command, 1773* (London).

Pieters, F.J.J.M. (2002). 'Het schatrijke naturaliënkabinet van Stadhouder Willem V onder directoraat van topverzamelaar Arnout Vosmaer', pp. 19-44 in B.C. Sliggers & M.H. Besselink, eds., *Het verdwenen museum: natuurhistorische verzamelingen 1750-1850* (Blaricum/Haarlem).

Quarrell, W.H. and W.J.C. Quarrell (1928). *Oxford in 1710 from the Travels of Zacharias Conrad von Uffenbach* (London).

Quarrell, W.H.. and Margaret Mare (1934). *London in 1710 from the Travels of Zacharias Conrad von Uffenbach* (London).

Ramsden (1774). *Description of a New Universal Equatoreal, made by Mr J. Ramdsen, with the Method of Adjusting it for Observation* (London). For further bibliographical details see McConnell 2007, pp. 279-280.

Reilly, Robin (1973). *Wedgwood Portrait Medallions. An Introduction* (London).

Reilly, Robin (1995). *Wedgwood. The New Illustrated Dictionary* (Woodbridge: Antique Collectors Club).

Rigaud, G. (1882). 'Dr Demainbray and the King's Observatory at Kew', *The Observatory* 5, pp. 279-285.

Rigaud, S.F.D. (1984). 'Facts and recollections of the XVIIIth century in a memoir of John Francis Rigaud', ed. W.L. Pressly, *Walpole Society* vol. 50, pp. 1-164.

Rooseboom, Maria (1950). *Bijdrage tot de geschiedenis der instrumentmakerskunst in de Noordelijke Nederlanden tot omstreeks 1850* (Leiden).

Scholten, Frits (2001). *Sumptuous memories: studies in 17ᵗʰ century Dutch tomb sculpture* (Zwolle).

Scott, Robert Henry (1885). 'The History of the Kew Observatory', *Proceedings of the Royal Society of London* 39, pp. 37-85.

Smith, Robert (1738). *A Compleat System of Opticks in four books, viz. a popular, a mathematical, a mechanical, and a philosophical Treatise: to which are added remarks upon the whole* (Cambridge, 2 vols.).

Sorrenson, Richard (2001). 'Dollond & Son's Pursuit of Achromaticity, 1758-1789', *History of Science* 39, pp. 31-55.

Steenstra, Pybo (1771-72). *Grondbeginsels der Sterrekunde* (Amsterdam, 2 vols).

Stern, W.M. (1969-70). 'Fish supplies for London in the 1760s: an experiment in overland transport', *Journal of the Royal Society of Arts*, 118, pp. 360-364 and 430-435.

Thykier, Claus, ed. (1990). *Dansk Astronomi Gennem Firehundrede Đr* (Copenhagen, 3 vols.)

Trueman Wood, Sir Henry (1913). *A History of the Royal Society of Arts* (London).

Turner, G. L'E. (1973). 'Descriptive Catalogue of Van Marum's Scientific Instruments in Teyler's Museum', in *Martinus van Marum. Life and Work* (6 vols., Haarlem and Leiden), vol. IV, pp. 127-396.

Turner, G. L'E. (1973a). 'A Very Scientific Century', in *Martinus van Marum. Life and Work* (6 vols., Haarlem and Leiden), vol. IV, pp. 3-38.

Turner, G. L'E. (1986). 'The Physical Sciences', in L.S. Sutherland and L.G. Mitchell, eds., *The History of the University of Oxford. Volume V: The Eighteenth Century* (Oxford 1986), pp. 659-681

Uffenbach, Z.C. von (1753-54). *Merkwürdige Reisen durch Niedersachsen, Holland und England* (3 vols., Ulm).

Uglow, Jenny (1997). *Hogarth. A Life and a World* (London).

Voorn, H. (1960). *De papiermolens in de provincie Noord-Holland* (Haarlem).

Vries, Jan de (1918), *Barges and capitalism. Passenger transport in the Dutch economy, 1632-1839* (Utrecht).

Walker, Annabel, with Peter Jackson, *Kensington and Chelsea. A social and architectural history* (London).

Wallis, Ruth D.C. Wallis, 'Cross-currents in Astronomy and Navigation: Thomas Hornsby, F.R.S. (1733-1810)', *Annals of Science* 57, pp. 219-240.

Wallis, R.V. and P.J. ([1986]. *Biobibliography of British Mathematics and its Applications Part II 1701-1760* (Newcastle upon Tyne).

Warner, Deborah Jean (1998). 'Edward Nairne: scientist and instrument-maker', *Rittenhouse* 12, pp. 65-93.

Watkins, Richard (2002). *Jérôme Lalande. Diary of a Trip to England 1763* (Kingston, Australia. On-line at http://www.watkinsr.id.au/lalande.html

Whinney, Margareth (1988). *Sculpture in Britain 1530 to 1830* (London, revised edition by John Physick, London).

Wijnman, H.F., ed. (1971). *Historische Gids van Amsterdam* (Amsterdam).

Wilson, Benjamin (1778). 'New experiments and observations on the nature and use of conductors', *Philosophical Transactions of the Royal Society of London* 68, pp. 245-313.

Witkam, H.J. (1980). *Catalogues of all the chiefest rarities in the Publick Anatomie Hall of the University of Leyden* ([Leiden])

Woolrich, A.P., ed. (1986). *Ferrner's journal 1759-1760 : an industrial spy in Bath and Bristol.* (Eindhoven : De Archaeologische Pers, [1986])

Your Time (2008). *Your Time. Including the Contribution of Northwest England to the Development of Clocks and Watches.* Published by the Antiquarian Horological Society as a catalogue for exhibitions at Prescot Museum and Williamson Museum & Art Gallery, Birkenhead, England, in 2008.

Zuidervaart, Huib J. (1999). *Van 'konstgenoten' en hemelse fenomenen: Nederlandse sterrenkunde in de achttiende eeuw* (Rotterdam).

Zuidervaart, Huib J. (2003). '"Zo'n mooie machine, waarvan de kwaliteit door alle astronomen wordt erkend". Een biografie van een vrijwel niet gebruikte telescoop', *Gewina* 26, pp. 148-165.

Zuidervaart, Huib J. (2004). 'Reflecting "Popular Culture". The Introduction, Diffusion and Construction of the Reflecting Telescope in the Netherlands', *Annals of Science* 61, pp. 407-452.

Zuidervaart, Huib J. (2006). 'Meest alle van best mahoniehout vervaardigd'. Het natuurfilosofisch instrumentenkabinet van de Doopsgezinde Kweekschool te Amsterdam, 1761-1828', *Gewina* 29, pp. 81-112.

Zuidervaart, Huib J. (2007). *Telescopes from Leiden Observatory and other collections 1656-1859, A Descriptive Catalogue* (Leiden: Museum Boerhaave).

Zuylen, J. van (1987). 'Jan en Harmanus van Deijl. Een optische werkplaats in de 18e eeuw', *Tijdschrift voor de Geschiedenis der Geneeskunde, Natuurwetenschappen, Wiskunde en Techniek* 10, pp. 208-228.

Photo credits

Figure page 18:
Photo: Museum Elburg.

Figure page 30:
Photo: Peter Louwman, Wassenaar.

Figure page 36:
Photo: Science and Society Picture Library, London.

Figure page 38:
Photo: Teylers Museum, Haarlem.

Figure page 44 left:
Photo: Teylers Museum, Haarlem.

Figure page 44 right:
Photo: Peter Louwman, Wassenaar.

Figure page 46:
Photo: Leiden Municipal Archive.

Figure page 47:
Photo: Museum Boerhaave, Leiden.

Figure page 48:
Photo: Museum Boerhaave, Leiden.

Figure page 51 left:
Photo: Museum Boerhaave, Leiden.

Figure page 51 right:
Photo: Teylers Museum, Haarlem.

Figure page 55:
Photo: The Hague Municipal Archives.

Figure page 64:
Photo: Bridgeman Art Library.

Figure page 72:
Photo: courtesy Dr J.T. van Doesburgh, Terwolde.

Figure page 74:
Photo: Science and Society Picture Library.

Figure page 79:
Photo: Science and Society Picture Library.

Figure page 80:
Photo: Science and Society Picture Library.

Figure page 81:
Photo: Museum of the History of Science, Oxford.

Figure page 84:
Photo: Science and Society Picture Library.

Figure page 89:
Photo: Teylers Museum, Haarlem.

Figure page 97:
Photo: Bridgeman Art Library.

Figure page 98:
Photo: Science and Society Picture Library.

Figure page 113:
Photo: Science and Society Picture Library.

Figure page 122:
Photo: Museum of the History of Science, Oxford.

Figure page 124:
Photo: Pitt Rivers Museum, Oxford.

Figure page 127:
Photo: Museum of the History of Science, Oxford.

Figure page 137:
Photo: Whipple Museum of the History of Science, Cambridge.

Figure page 139:
Photo: St John's College, Cambridge.

Figure page 142:
Photo: Science and Society Picture Library.

Figure page 149:
Photo: Science and Society Picture Library.

Figure page 165:
Photo: British Museum, London.

Figures page 166:
Photo: © National Maritime Museum, Greenwich, London.

Figure page 167:
Photo: The Round Tower Museum, Copenhagen.

Name index

Subject index